PHYSICS RESEARCH AND TECHNOLOGY

MANGANITE-BASED THIN FILMS AND HETEROJUNCTIONS

PHYSICS RESEARCH AND TECHNOLOGY

PHYSICS RESEARCH AND TECHNOLOGY

MANGANITE-BASED THIN FILMS AND HETEROJUNCTIONS

ZHIGAO SHENG,
LING HU
AND
YUPING SUN
EDITORS

nova
publishers
New York

Library of Congress Cataloging-in-Publication Data

Manganite-based thin films and heterojunctions / editors: Zhigao Sheng, Ling Hu, and Yuping Sun (High Magnetic Field Laboratory, Key Laboratory of Materials Physics, Institute of Solid State Physics, Chinese Academy of Sciences, Anhui, P.R. China).
pages cm -- (Physics research and technology)
Includes index.
ISBN 978-1-63482-452-1 (hardcover)
1. Condensed matter. 2. Solids--Electric properties. I. Sheng, Zhigao, editor. II. Hu, Ling (Researcher on physics), editor. III. Sun, Yuping, 1964- editor. IV. Series: Physics research and technology.
QC173.458.E43M36 2015
546'.5415--dc23
2015008468

Published by Nova Science Publishers, Inc. † New York

CONTENTS

PREFACE

As a research focus of science, mixed-valence manganites started receiving considerable attention at the turn of the late 20th and the early 21st century. The intriguing interplay of charge, spin and orbital ordering in the manganites systems superimposed by interface/surface effects provides a scientific platform beneficial for both fundamental and application-oriented researches. The fundamental basis for understanding physics of manganites is the concept of the symmetry of order parameters which can be manipulated artificially by controlling the interplay of electronic degrees of freedom. From this viewpoint, the possibility of externally modifying the properties of manganite based thin films and heterojunctions by epitaxial strain or artificial boundaries, thus generating new properties at the interface between film and heterointerface, opens a new perspective. This book is intended primarily to enable the comprehensive understanding of manganite-based thin films and heterojunctions, which is in dire need.

This book deals with the research progresses of manganite-based films and heterojunctions. Chapter 1 presents an overview of manganite films and heterojunctions. The strain induced effects in manganite-based films and superlattices are discussed in Chapter 2. Chapter 3 treats the electrical transport in manganite based heterojunctions. Chapter 4 offers an introduction of the photoinduced and photovoltaic effect in manganite based films and heterojunctions. The last Chapter, Chapter 5, is an outline of magneto-tunable effects of manganite based heterojunctions.

The authors want to thank all his colleagues and students who helped in the preparation of the book. This includes Qin Li, who helped in the editing and proofreading job. Thanks are also due to Sisi Chen, Jingjing Gao, Guangyuan He and Zhehan Li for their devotion to the preparation of the references.

In: Manganite-Based Thin Films ... ISBN: 978-1-63482-452-1
Editors: Z. Sheng, L. Hu and Y. Sun

Chapter 1

OVERVIEW OF MANGANITE

L. Hu[1], *Z. G. Sheng*[2,*] *and Y. P. Sun*[1,2]
[1]Key Laboratory of Materials Physics,
Institute of Solid State Physics, Chinese Academy of Sciences,
Hefei, People's Republic of China
[2]High Magnetic Field Laboratory, Chinese Academy of Sciences,
Hefei, People's Republic of China

Abstract

Doped manganites with a perovskite structure are of great interest due to their rich physical properties and potential applications. The manganites exhibit various kinds of electronic phases such as charge- and orbital-ordered insulator, ferromagnetic metal, ferromagnetic insulator, and paramagnetic insulator. These competing phases originate from the strongly correlated charge, spin, orbital, and lattice degrees of freedom and produce complex phase diagrams. This chapter is intended to offer an introduction to some fundamental and key experiment results of the perovskite manganites. It deals with the crystal structure, electronic structure, phase diagram and rich physical properties of these compounds, manganite-based thin film and heterojunction.

PACS: 75.47.Lx, 75.47.Gk, 71.27.+a, 71.30.+h

Keywords: manganite, phase separation

---

*E-mail address: zhigaosheng@hmfl.ac.cn

1. Introduction

Perovskite transition metal oxides with ABO_3 stoichiometry are a group of materials that exhibit many interesting physical properties and functional behaviors such as ferroelectricity, superconductivity, colossal magnetoresistance (CMR), and metal-insulator transition. Among them, the terminology CMR was coined to denote the gigantic negative magnetoresistance observed in the perovskite manganites. Historically, the mixed-valence perovskites $La_{1-x}A_xMnO_3$ (A=Ca,Sr,Ba) were studied in the early 1950s, both experimentally and theoretically [1, 2, 3, 4, 5, 6]. The experiments performed on polycrystalline samples showed antiferromagnetic (AFM) insulating behaviour at low and high x values and ferromagnetic (FM) metallic behaviour in a certain range of concentrations centred around x≈1/3 [1, 2]. This results triggered the development of the double exchange (DE) mechanism [3, 4, 5, 6]. Zener considered that the intra-atomic Hund coupling was strong and that the carriers do not change their spin orientation when hopping from one ion to another, so they can only hop if the spins of the two ions are parallel. On minimizing the total free energy of the system, it was found that FM interactions are favoured when the magnetic atoms are fairly well separated and conduction electrons are present. Starting from the AFM insulating $LaMnO_3$ end member, Zener showed how the system gradually become more FM upon hole doping (introduction of Mn^{4+}). He considered the problem of the exchange between Mn^{3+} and Mn^{4+} ions via an oxygen ion and introduced the concept of simultaneous transfer of an electron from the Mn^{3+} to the oxygen and from the oxygen to the neighbouring Mn^{4+} (Fig. 1.1). Such transfer is called DE. DE means a magnetic interaction mediated by itinerant spin-polarized 3d electrons which couple into according to Hund's rule the localized magnetic moments. Itinerancy of electrons is energetically favourable for kinetic energy gain. In a semiclassical treatment, DE leads to a specific dependence of electron transfer between Mn ions at an angle of Θ between their magnetic moments: $t = t_0 \cos(\frac{\Theta}{2})$. The transfer probability varies from 1 for $\Theta = 0$ to zero for $\Theta = 180^o$. This simple relation illustrates the origin of a large MR observable in DE ferromagnets: if a magnetic field is strong enough to align the magnetic core moments, resistance would drop drastically.

Interest in the mixed-valence manganites was revived in the 1990s following the preparation of high-quality manganite thin films with large magnetoresistance [7, 8]. The manganite thin films showed remarkable magnetoresistance effects which were epitomized by Jin *et al.* as the 'colossal magnetoresistance' (CMR). The Curie point coincides with a metal-insulator transition, and there are relative anomalies in various physical properties. Since then, a lot of works were devoted to the mixed-valence oxides, mainly to $R_{1-x}A_xMnO_3$ (R=rare earth, Y, and Bi, A=divalent alkaline earth) with three-dimensional perovskite structure. The system offers chemical flexibility which enables the systematic research into the relation among the perovskite structure, electronic and magnetic properties.

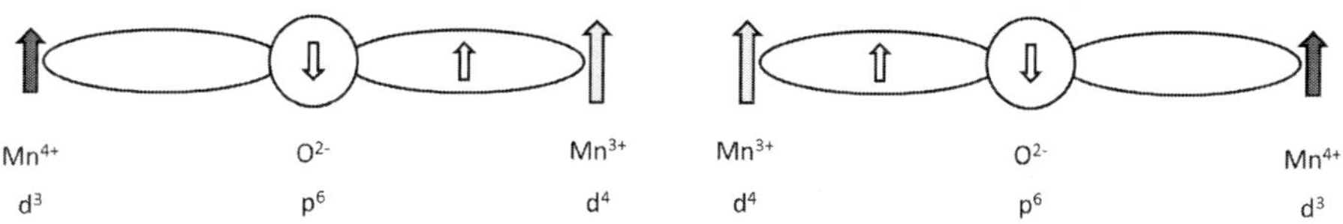

Figure 1.1. Schematic diagram of the DE mechanism.

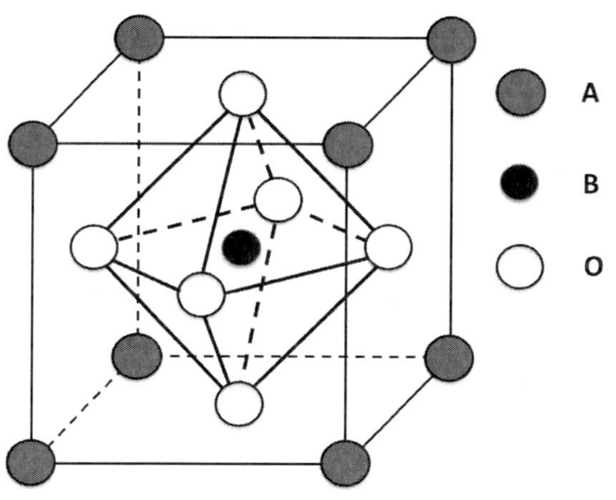

Figure 1.2. Schematic view of the cubic perovskite structure.

2. Crystallographic Structure and Electronic Structure

The structure of $R_{1-x}A_xMnO_3$ is close to that of the cubic perovskite (Fig. 1.2). The large sized R and A ions occupy the A-site with 12-fold oxygen coor-

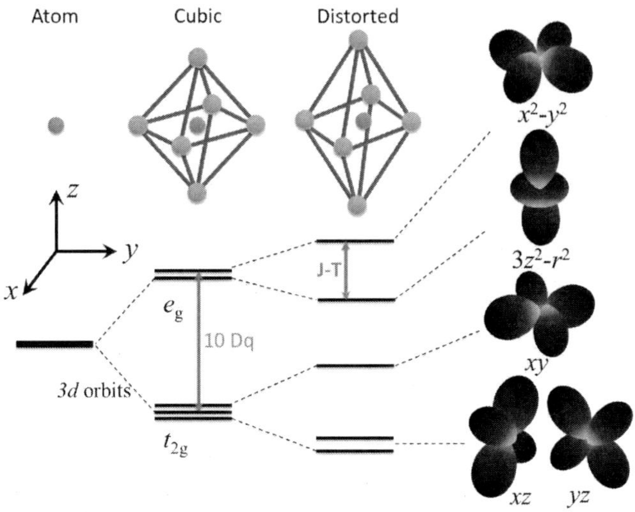

Figure 1.3. Crystal-field splitting of the five-fold degenerate atomic 3d levels into lower t_{2g} (triply degenerate) and higher e_g (doubly degenerate) levels.

dination. The smaller Mn ion are located at the centre of the oxygen octahedron with 6-fold coordination. Perovskite-based structures occasionally show lattice distortion from the cubic structure. In these distorted perovskites, the MnO_6 octahedra show alternating buckling. Such a lattice distortion of the perovskite in the $R_{1-x}A_xMnO_3$ is governed by the so-called tolerance factor t, which is defined as

$$t = (r_A + r_O)/\sqrt{2}(r_B + r_O) \tag{1}$$

r_A and r_B denote the average ionic radii at the A and B sites and r_O is the radius of the O^{2+} ion. The tolerance factor t measures, by definition, the lattice-matching of the sequential AO and BO_2 planes. When t is close to 1, the cubic perovskite structure is realized. As r_A or equivalently t decreases, the lattice structure transforms to the rhombohedral ($0.96 < t < 1$) and then to the orthorhombic structure ($f < 0.96$), in which the Mn-O-Mn bond angle (Θ) is bent and deviates from 180°. Most manganites represent the case $t < 1$.

For an isolated 3d ion, five degenerated orbital states are available to the 3d electrons with $l = 2$. In a crystal, the degeneracy is partly lifted by the crystal field. The five 3d-orbitals are split by a cubic crystal field into three t_{2g} orbitals and two e_g orbitals. For the manganites, the Mn^{3+} shows the electronic configuration of $t_{2g}^3 e_g^1$ (S=2)(Fig. 1.3). This is due to the Hund's coupling energy J_H is larger than the crystal field splitting energy $10D_q$. In a crystal field of symmetry lower than cubic, the degeneracy of the e_g and t_{2g} levels is lifted. Furthermore, the Mn^{3+} has a marked tendency to distort its octahedral environment to lift the degeneracy of e_g level and lower the overall energy (Jahn-Teller distortion). The most frequently observed Jahn-Teller distortion is, for example, the elongation of the z-axis oxygen positions coupled with the $d_{3z^2-r^2}$ orbital, or the elongation of the x and y-axes (in-plane) oxygen positions coupled with the $d_{x^2-y^2}$ orbital. This Jahn-Teller distortion is rather effective in the lightly doped manganites, i.e., with a large concentration of Mn^{3+} ions. In this case, the individual Jahn-Teller distortions are cooperatively induced, leading to symmetry lowering of the lattice form or a new superstructure. As a result, the obital ordering (OO) in the manganites is always associated with the cooperative Jahn-Teller distortion. The Jahn-Teller distortion is also the source of strong electron-phonon coupling. In doped manganites, superstructures of localized charge carriers (charge order, CO) usually accompany the orbital order. CO/OO has been observed for numerous insulating manganites. Moving e_g electrons may form lattice polarons arising from the Jahn-Teller distortion, and the trapping (localization) of charge carriers by lattice distortion is possible. For metallic manganites, the Jahn-Teller distortions are dynamic, resulting in an average distortion of all MnO_6 octahedra.

Another important feature in the perovskite is that the compound is quite suitable for the carrier doping procedure (filling control) since the structure is very robust against the chemical modification on the A-site such as the solid solution of the divalent (alkaline-earth) and trivalent (rare-earth) ions. In the $R_{1-x}A_xMnO_3$, the averaged Mn valence varies as 3+x. In other words, x produces a vacancy in the e_g electron state at a rate of x per Mn-site and is referred to as hole-doping. When electron vacancies or holes are created in the e_g orbital states of the crystal and escape from the long-range ordering, the e_g electrons can be itinerant and hence play the role of conduction electron.

3. Phase Diagram of Manganites

The important contributions to electronic energy in manganites are summarized as follows: (i) kinetic energy of e_g electrons, (ii) the Hund on-site magnetic coupling between e_g and t_{2g} electron spins, (iii) coupling of e_g electrons and elastic distortions of MnO_6 octahedra (Jahn-Teller effect), (iv) the crystal field split-

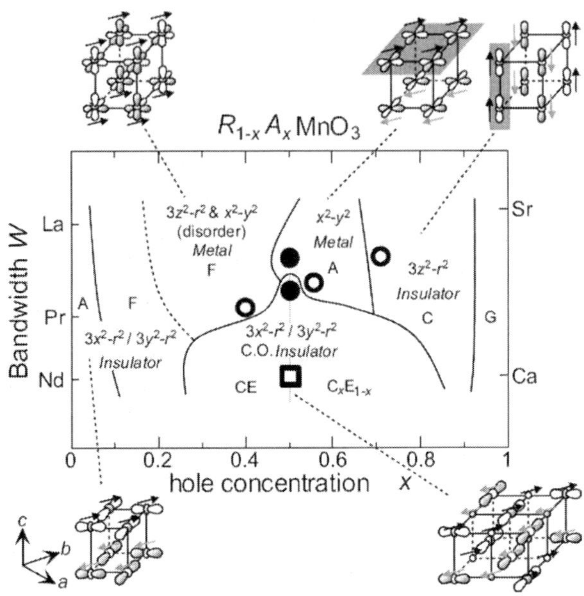

Figure 1.4. Bandwidth versus hole concentration phase diagram at the ground state of $R_{1-x}A_xMnO_3$. F denotes a FM state. A, CE, C and G denote A-type, CE-type, C-type and G-type AFM states, respectively. Illustrations show schematic views of the orbital and spin configurations on Mn sites in the respective phases(©IOP Publishing. Reproduced with permission from Reference [9]. All rights reserved).

ting energy in octahedral coordination, (v) the Heisenberg magnetic coupling between the nearest-neighbour localized electron spins and (vi) the Coulomb interaction among e_g electrons. The dominance of one or other mechanism results in various ground states. The ground state of $R_{1-x}A_xMnO_3$ depends on (i) the number of doped charge carriers x, (ii) the average ionic radii on La and Mn lattice sites and (iii) the scatter of ionic radii on La site measured by a disorder parameter σ. Ground states observed for manganites include ferromagnetic metals (FMM), AFM insulators, ferromagnetic insulators (FMI), AFM metals, glassy insulators, more complex canted magnetic structures and mixed-phase states (Fig. 1.4) [9]. When the bandwidth is reduced with the decrease in the tolerance factor, the sequential A-F-A-C transformation of the long-range magnetic order with hole-doping is violated by the subsistence of the CO/OO (the CE-type spin order)(Fig. 1.5).

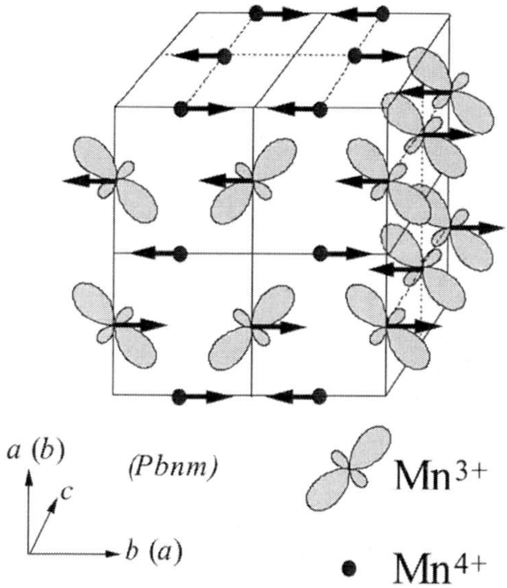

Figure 1.5. Spin, charge and e_g orbital ordering pattern of the CE type observed for most of the x ≈1/2 manganites. Lobes on the nominal Mn^{3+} sites represent the e_g orbital. The Mn^{4+} sites are indicated by closed circles(©IOP Publishing. Reproduced with permission from Reference [9]. All rights reserved).

Phase diagrams in dependence on temperature and doping level x can be illustrated by three prominent examples: $La_{1-x}Sr_xMnO_3$, $Nd_{1-x}Sr_xMnO_3$, and $Pr_{1-x}Ca_xMnO_3$ [10]. The ionic radius r_A or equivalently the tolerance factor t decreases from La-Sr system to Pr-Ca through Nd-Sr, leading to increasing distortions of the lattice and decreasing in the one-electron bandwidth (W). This means that other electronic instabilities, such as CO/OO and super-exchange interactions which compete with the DE interaction, may become dominant in specific regions of x and temperature.

$La_{1-x}Sr_xMnO_3$ has the maximal W. The end compound $LaMnO_3$ (x=0) is strongly affected by both the collective Jahn-Teller effect and the electron correlation effect. The collective Jahn-Teller distortion present in $LaMnO_3$ reflects the OO such as the alternating $d_{3x^2-r^2}$ and $d_{3y^2-r^2}$ orbitals on the ab plane. In this Jahn-Teller-distorted and OO state, the $LaMnO_3$ undergoes an AFM transition at $T_N = 140K$, which is much lower than the OO temperature (780 K)[11]. The spin ordering structure is the layered type or the so-called A-type, in which the ferromagnetic ab plane is coupled antiferromagnetically along the c-axis.

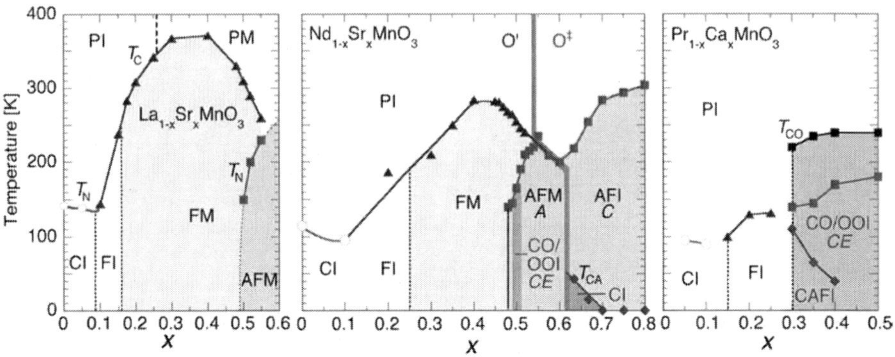

Figure 1.6. Magnetic and electronic phase diagram versus doping x for $La_{1-x}Sr_xMnO_3$ (large bandwidth), $Nd_{1-x}Sr_xMnO_3$ (medium bandwidth), $Pr_{1-x}Ca_xMnO_3$ (low bandwidth) (Reprinted from Reference [10], Copyright (1999), with permission from Elsevier).

With hole-doping by substitution of La with Sr, the ordered spins are more canted towards the c-axis direction. De Gennes interpreted that the extension of the wavefunction of the hole along the c-axis can mediate the DE-type ferromagnetic coupling, producing the spin canting [6]. As the doping level x increases typically above 0.1, the spin state is turned into a ferromagnetic one, while the compound remains insulating or semiconducting up to the critical doping level for the metal-insulator transition, e.g., x = 0.17 [12]. The consequence of hole-doping is that the static collective Jahn-Teller distortion present in $LaMnO_3$ is diminished with increase in the doping level. The nature of the FMI state for $0.1 < x < 0.17$ in $La_{1-x}Sr_xMnO_3$ has been confirmed as the CO/OO state [13, 14, 15, 16, 17]. It is believed that the FM super-exchange interaction under strong electron correlation produces the three-dimensional FM state with insulating behaviour [13, 14]. With doping above x=0.17, the FMM phase appears below T_C which steeply increases with x up to 0.3 and then saturates. The half-metallic nature of the FMM phase of $La_{0.7}Sr_{0.3}MnO_3$ has been demonstrated by spin-resolved photoemission measurements and Andreev reflection [18, 19]. The $La_{0.7}Sr_{0.3}MnO_3$ (T_C ~370 K) exhibits potential applications for the spintronic devices. With further hole-doping above x=0.5, the metallic but AFM phase appears. The AFM phase for $x > 0.5$ is the A-type and the metallic state is accompanied by a uniform ordering of the $d_{3x^2-r^2}$ orbital. The metallic

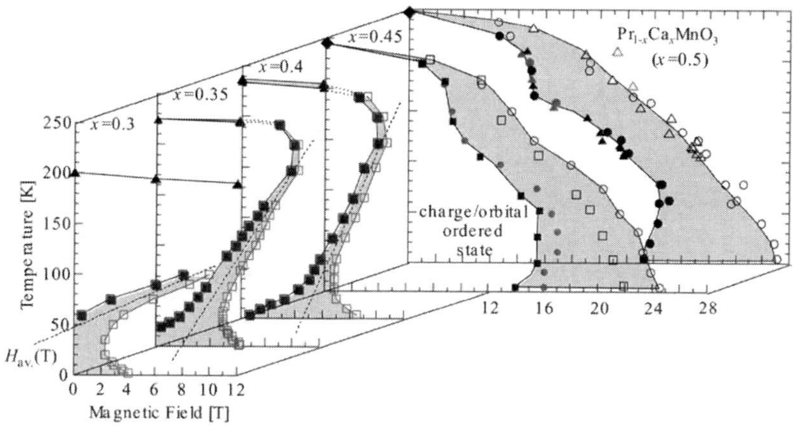

Figure 1.7. The charge/orbital-ordered state of $Pr_{1-x}Ca_xMnO_3$ (x=0.5, 0.45, 0.4, 0.35 and 0.3), which is plotted on the magnetic field-temperature plane. The hatched area indicates the hysteresis region(Reprinted from Reference [10], Copyright (1999), with permission from Elsevier).

$d_{x^2-y^2}$ electron band is formed due to the double-exchange interaction, while the FM sheet is coupled antiferromagnetically along the c-axis mainly due to the super-exchange interaction of the t_{2g} electron spins. This (quasi-)metallic $d_{x^2-y^2}$ orbital-ordered state is ubiquitous in the overdoped (x=0.5-0.6) manganites with relatively large W, e.g., $(Pr,Nd)_{1-x}Sr_xMnO_3$ [20, 21].

$Pr_{1-x}Ca_xMnO_3$ has low W and the FMM state is no longer realized at zero magnetic field and at ambient pressure. The FMI phase appears for $0.15 < x < 0.3$. With further increase of the hole-doping level, the CO/OO states with 1:1 ordering of Mn^{3+}/Mn^{3+} appears for $x > 0.3$. The compound exhibits a transition from a paramagnetic insulating phase (PMI) to a CO/OO state at relatively high temperatures in a wide doping range $0.3 \leqslant x \leqslant 0.75$. The CO/OO pattern $Pr_{1-x}Ca_xMnO_3$ is shown in Fig. 1.5. The $d_{3x^2-r^2}$ and $d_{3y^2-r^2}$ orbitals form the stripes along the diagonal of the Mn-square lattice. The $Pr_{1-x}Ca_xMnO_3$ compounds undergo CO/OO transition at higher temperature than the long-rang CE-type spin ordering transition. In an applied field which aligns the Mn spins, a transition to a FMM phase is observed with a resistivity change by several orders of magnitude (Fig. 1.7). The phase boundaries have been determined by the measurements of the magnetic field dependence of resistivity (ρat-H) and magnetization (M-H) at fixed temperatures. The hatched area in the figure represents the magnetic-field hysteresis region accompanied by the first-order nature of the transition. This first-order transition is accompanied by abrupt changes in lattice parameters and large hysteresis effects. The critical field to destroy the CO/OO state at x=0.5 becomes as large as about 27 T at 4.2 K. In the case of x=0.3, collapse of the CO/OO state is realized by applying an external magnetic field of only several tesla when the temperature is set below 50K. Besides magnetic field, the CO/OO state can be destroyed by other external stimuli such as electric field, x-ray, and photo irradiation [22, 23, 24, 25].

In contrast, $Nd_{1-x}Sr_xMnO_3$ has intermediate W and exhibits complicate phase diagram. The FMM phase appears if $0.25 < x < 0.5$. when the doping x is close to the commensurate value x=1/2, the CO state comes to the surface. The FMM phase of $Nd_{1/2}Sr_{1/2}MnO_3$ changes into the CO/OO insulating state at $T_{CO/OO}$=160 K when decreasing the temperature [26]. This CO/OO state accompanies the concomitant AFM ordering of spins, which is called the CE type (Fig. 1.5). The nominal Mn^{3+} and Mn^{4+} species in a ratio of 1:1 ratio show real-space ordering on the (001) plane of the orthorhombic lattice , while the e_g orbital shows the 1×2 superlattice on the same (001) plane. The CE-type charge/spin order formed in this system is limited in a very narrow region of x around 1/2. With the hole concentration of $x > 0.52$, the A-type AFM ordering turns up. The uniform ordering of $d_{x^2-y^2}$ orbitals makes the phase metallic but highly 2D-like. This is because the inter-plane (c-axis) hopping of the fully spin-polarized $d_{x^2-y^2}$ electron is almost forbidden due to the in-plane confinement in the layered-AF structure as well as the minimal supertransfer interaction between the $d_{x^2-y^2}$ orbitals along the z-direction. As x is further increased, the rod type (C-type) AFM ordering turns up at $x > 0.62$. the $d_{3z^2-r^2}$ orbitals are mostly occupied in the C-type AFM phase (Fig. 1.4).

Besides the hole-doping level x, the one-electron bandwidth W can solely be used to control the electronic phase in manganites. The W of manganites are controlled by the averaged radius of A-site ions. If the averaged radius decreases, the tilting of MnO_6 octahedra increases in an orthorhombically distorted structure. Then the hybridization between the manganese e_g and oxygen 2p states decreases or the W reduces. In the reduced-W cases, other interactions competing with the DE interaction become relatively important, such as the repulsive Coulomb interaction among the e_g electrons, the CO/OO, the superexchange interaction between the local t_{2g} spins, the collective Jahn-Teller distortion, etc. One typical example is $Pr_{0.55}(Ca_{1-y}Sr_y)_{0.45}MnO_3$, in which the

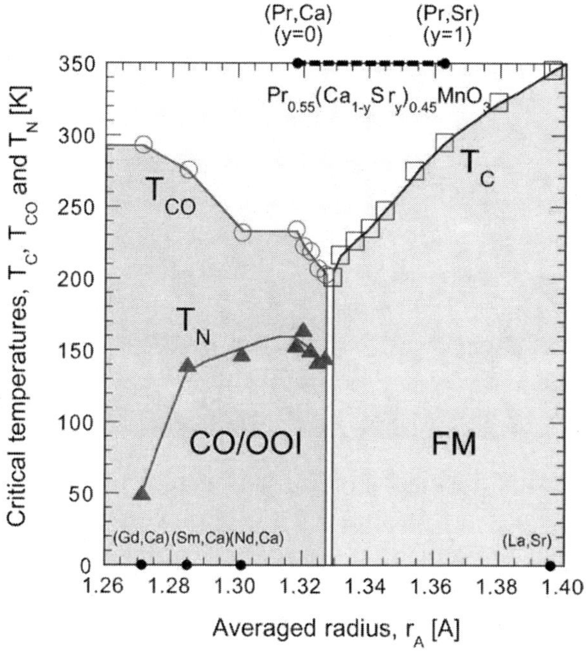

Figure 1.8. The extended phase diagram for x=0.45 over a wide regime of the A-site averaged ionic radius(Reprinted with permission from Reference [27], Copyright (2002) by the American Physical Society).

isovalent substitution of Ca with Sr gradually increase the W without changing the doping x. The end members of the series, $Pr_{0.55}Ca_{0.45}MnO_3$ (y=0) and $Pr_{0.55}Sr_{0.45}MnO_3$ (y=1) have robust low-temperature CO/OO and FMM states, respectively. Upon Sr doping, the competition between the CO/OO and FMM exhibits bicritical feature: both $T_{CO/OO}$ and T_C systematical decrease towards the CO/OO-FMM phase boundary and finally coincide with each other at y=0.25, forming the bicritical point (Fig. 1.8)[27]. Such phase competition is prevalent in the doped manganites and $La_{0.625-y}Pr_yCa_{0.375}MnO_3$ is another typical model system of phase competition among different electronic states. The ground states of $La_{0.625}Ca_{0.375}MnO_3$ (y=0) and $Pr_{0.625}Ca_{0.375}MnO_3$ (y=1) are FMM and CO/OO states, respectively. The phase competition and coexistence of FMM and COOI states appear at intermediate Pr contents of y=0.3-0.4 [28].

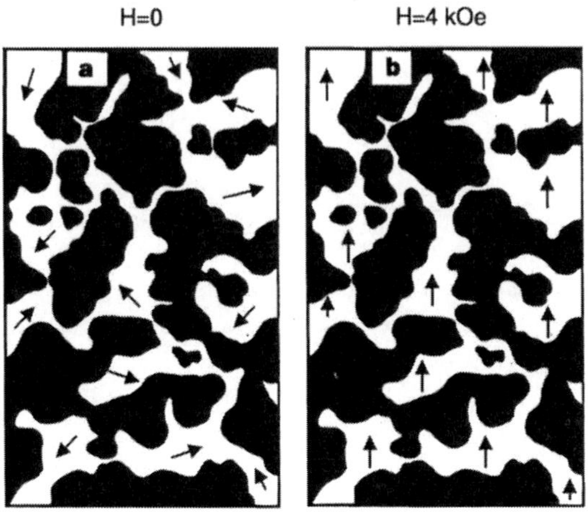

Figure 1.9. Schematic illustration of the sub-micrometre-scale coexistence of the CO/OO insulating (dark area) and FMM (white area) domains(Reprinted by permission from Macmillan Publishers Ltd: Nature (Ref.[28]), copyright (1999)) .

4. Electronic Phase Separation

Theoretical and experimental studies suggest that the ground states of manganite tend to be intrinsically inhomogeneous due to the presence of strong tendencies toward phase separation, typically involving FMM and AFM CO/OO insulating domains. The phase separation tendencies have two origins: (i) electronic phase separation between phases with different densities that lead to nanometer scale coexisting clusters, and (ii) disorder-induced phase separation with percolative characteristics between equal-density phases, driven by disorder near first-order metal-insulator transitions. The phase separation phenomenon occurs on various time-scales (from static to dynamic) and on various length-scales (from glass-like nano to grain-like micron) [29]. Experimental evidences for the electronic phase separation in manganites have been collected in many ways [28, 30, 31, 32, 33, 34, 35, 36, 37, 38]. Uehara *et al.* have demonstrated that the (La,Pr,Ca)MnO_3 system is electronically phase-separated into a sub-micrometre-scale mixture of CO/OO insulating and FMM domains using electron microscopy data (Fig. 1.9) [28]. Fath *et al.* have used scanning tunneling spectroscopy to investigate single crystals and thin films of $La_{1-x}Ca_xMnO_3$(with x of about 0.3), which exhibit CMR. The different spectroscopic signatures of the insulating (paramagnetic) and metallic (ferromagnetic) phases enable their spatial extent to be imaged down to a lateral scale of the order of 10 nanometers. Above the bulk transition temperature T_C, the images show mostly insulating behavior. Below T_C, a phase separation is observed where inhomogeneous structures of metallic and more insulating areas coexist and are strongly field dependent in their size and structure. Insulating areas are found to persist far below T_C. These results suggest that the transition and the associated magnetoresistance behavior should be viewed as a percolation of metallic ferromagnetic domains [31]. Zhang *et al.* have directly observed the formation of these percolative networks by directly measuring the temperature-dependent local magnetic microstructure in $La_{0.33}Pr_{0.34}Ca_{0.33}MnO_3$ thin films using a low-temperature magnetic force microscope [33]. Sheng *et al.* have studied the phase separation of CMR manganite thin films $Pr_{0.5}Ca_{0.5}MnO_3$ and $Pr_{0.55}(Ca_{0.75}Sr_{0.25})_{0.45}MnO_3$ using dielectric spectroscopy with unique junction configuration. It was found that the depletion layer in Nb-doped $SrTiO_3$ can act as isolation layer for exploration of dielectric properties of manganite thin films. The multiple dielectric relaxations have been detected in the manganite thin film with the phase separation between the COOI and FMM states (Fig. 1.10). The dynamics of multiple phases and their boundaries have been deduced separately from temperature- and magnetic-field-dependent variations of the dynamic capacitance spectra [38].

5. Manganite-based Thin Film and Heterojuction

Due to the rich physical properties of the CMR manganites, they have attracted much interest and have triggered intensive research towards potential applications, e.g., magnetic reading heads, magnetic random access memories, magnetic field sensors, magnetic refrigeration, and cathodes for solid oxide fuel cells [39, 40, 41, 42, 43, 44]. Since most applications require thin films with good magnetic and electrical properties, it is of great importance to be able to prepare high quality single-crystal epitaxial films. On the other hand, thin films are also important for fundamental research: they allow us to study the effects of finite thickness, biaxial strain, interface electronic structure and proximity effects, defects formed during growth. Since thin film growth is far from equilibrium conditions, new phases can be obtained. Furthermore, heterostructure

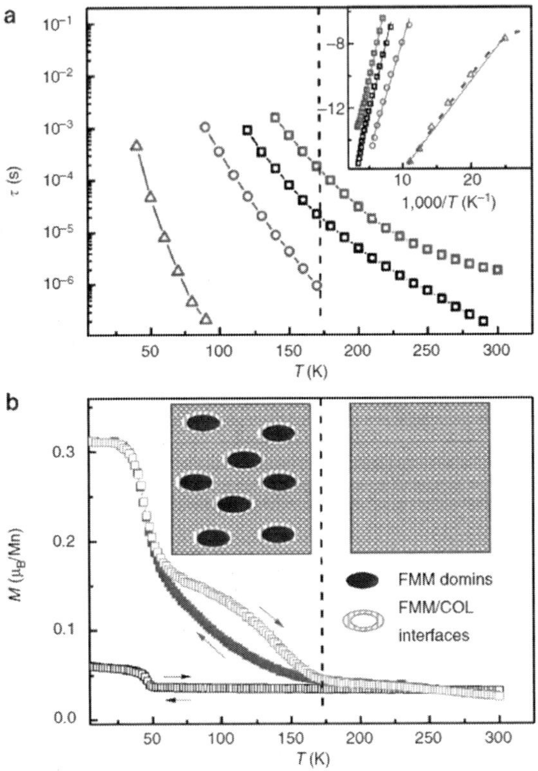

Figure 1.10. (a)Temperature dependence of relaxation time for PCMO (black) and PCSMO (red). The red curves with squares, circles, and triangles present data for τ_1, τ_2, τ_3 of PCSMO, respectively. The inset shows Arrhenius plot of relaxation times. The blue dotted line is fitting result for τ_3. (b)Temperature dependence of magnetization for PCSMO (black)and PCSMO (red) films measured under $\mu_0 H$=0.05 T. Arrows indicate the temperature scan directions. The insets schematically represent the coexistence of multiple phases in PCSMO sample. The solid ellipses and vertically patterned rings stand for FMM domains and interfaces between FMM and COOI domains, respectively(Reprinted by permission from Macmillan Publishers Ltd: Nature Communications (Ref.[38]), copyright (2012)).

formation is an effective approach to achieve reversible modulation of carrier concentration using electric field or light, and is widely used for semiconductor devices. Compared with ordinary semiconductors, the manganites exhibit some unique features because of the nature of strong electron correlation and the magnetic state dependent band structure. Therefore, it is of interest to investigate the properties of the manganite-based heterojunction in electric field, magnetic field and light aiming at the exploration of novel physical phenomena and theory.

References

[1] G. H. Jonker, and J. H. Van Santen, *Physica* 16, 337 (1950); J. H. Van Santen, and G. H. Jonker, *ibid* 16, 599 (1950); J. Volger, *ibid* 20, 49 (1954); G. H. Jonker, *ibid* 20, 118 (1954).

[2] E. O. Wollan, and W. C. Koehler, *Phys. Rev.* 100, 545 (1955).

[3] C. Zener, *Phys. Rev.* 81, 440 (1951); *ibid* 82, 403 (1951).

[4] J. B. Goodenough, *Phys. Rev.* 100, 564 (1955).

[5] P. W. Anderson, and H. Hasegawa *Phys. Rev.* 100, 675 (1955).

[6] P. G. de Gennes, *Phys. Rev.* 118, 141 (1960).

[7] R. von Helmolt, J. Wecker, B. Holzapfel, L. Schultz, and K. Samwer, *Phys. Rev. Lett.* 71, 2331 (1993).

[8] S. Jin, T. H. Tiefel, M. McCormack, R. A. Fastnacht, R. Ramesh, and L. H. Chen, *Science* 264, 413 (1994).

[9] Y. Tokura, *Rep. Prog. Phys.* 69, 797 (2006).

[10] Y. Tokura, and Y. Tomioka, *J. Magn. Magn. Mater* 200, 1 (1999).

[11] Y. Murakami, J. P. Hill, D. Gibbs, M. Blume, I. Koyama, M. Tanaka, H. Kawata, T. Arima, Y. Tokura, K. Hirota, and Y. Endoh, *Phys. Rev. Lett.* 81, 582 (1998).

[12] A. Urushibara, Y. Moritomo, T. Arima, A. Asamitsu, G. Kido, and Y. Tokura, *Phys. Rev. B* 51, 14103 (1995).

[13] Y. Endoh, K. Hirota, S. Ishihara, S. Okamoto, Y. Murakami, A. Nishizawa, T. Fukuda, H. Kimura, H. Nojiri, K. Kaneko, and S. Maekawa, *Phys. Rev. Lett.* 82, 4328 (1999).

[14] H. Nojiri, K. Kaneko, M. Motokawa, K. Hirota, Y. Endoh, and K. Takahashi, *Phys. Rev. B* 60, 4142 (1999).

[15] S. Uhlenbruck, R. Teipen, R. Klingeler, B. Buchner, O. Friedt, M. Hucker, H. Kierspel, T. Niemoller, L. Pinsard, A. Revcolevschi, and R. Gross, *Phys. Rev. Lett.* 82, 185 (1999).

[16] F. Moussa, M. Hennion, F. Wang, P. Kober, J. Rodriguez-Carvajal, P. Reutler, L. Pinsard, and A. Revcolevschi, *Phys. Rev. B* 67, 214430 (2003).

[17] G. Papavassiliou, M. Pissas, G. Diamantopoulos, M. Belesi, M. Fardis, D. Stamopoulos, A. G. Kontos, M. Hennion, J. Dolinsek, J.-Ph. Ansermet, and C. Dimitropoulos, *Phys. Rev. Lett.* 96, 097201 (2006).

[18] J.-H. Park, E. Vescovo, H.-J. Kim, C. Kwon, R. Ramesh, and T. Venkatesan, *Nature* 392, 742 (1998).

[19] R. J. Soulen Jr., J. M. Byers, M. S. Osofsky, B. Nadgorny, T. Ambrose, S. F. Cheng, P. R. Broussard, C. T. Tanaka, J. Nowak, J. S. Moodera, A. Barry, J. M. D. Coey, *Science* 282, 85 (1998).

[20] Y. Moritomo, T. Akimoto, A. Nakamura, K. Ohoyama, and M. Ohashi, *Phys. Rev. B* 58, 5544 (1998).

[21] H. Kuwahara, T. Okuda, Y. Tomioka, A. Asamitsu, and Y. Tokura, *Phys. Rev. Lett.* 82, 4312 (1999).

[22] V. Kiryukhin, D. Casa, J. P. Hill, B. Keimer, A. Vigiante, Y. Tomioka, and Y. Tokura, *Nature* 386, 813 (1997).

[23] A. Asamitsu, Y. Tomioka, H. Kuwahara, and Y. Tokura, *Nature* 388, 50 (1997).

[24] M. Fiebig, K. Miyano, Y. Tomioka, and Y. Tokura, *Science* 280, 1925 (1998).

[25] M. Rini, R. Tobey, N. Dean, J. Itatani, Y. Tomioka, Y. Tokura, R. W. Schoenlein, and A. Cavalleri, *Nature* 449, 72 (2007).

[26] H. Kuwahara, Y. Tomioka, A. Asamitsu, Y. Moritomo, Y. Tokura, *Science* 270, 961 (1995).

[27] Y. Tomioka and Y. Tokura, *Phys. Rev. B* 66, 104416 (2002).

[28] M. Uehara, S. Mori, C.H. Chen, and S.-W. Cheng, *Nature* 399, 560 (1999).

[29] E.Dagotto, T. Hotta, and A. Moreo, *Phys. Rep.* 344, 1 (2001).

[30] C. H. Booth, F. Bridges, G. H. Kwei, J. M. Lawrence, A. L. Cornelius, J. J. Neumeier, *Phys. Rev. Lett.* 80, 853 (1998).

[31] M. Fath, S. Freisem, A. A. Menovsky, Y. Tomioka, J. Aarts, and J. A. Mydosh, *Science* 285, 1540 (1999).

[32] P. Dai, J. A. Fernandez-Baca, N. Wakabayashi, E. W. Plummer, Y. Tomioka, Y. Tokura, *Phys. Rev. Lett.* 85, 2553 (2000).

[33] L. W. Zhang, C. Israel, A. Biswas, R. L. Greene, A. de Lozanne, *Science* 298, 805 (2002).

[34] S. Yoon, H. L. Liu, G. Schollerer, S. L. Cooper, P. D. Han, D. A. Payne, S. W. Cheong, Z. Fisk, *Phys. Rev. B* 58, 2795 (1998).

[35] V. Chechersky, A. Nath, C. Michel, M. Hervieu, K. Ghosh, R. L. Greene, *Phys. Rev. B* 62, 5316 (2000).

[36] Y. Q. Ma, W. H. Song, R. L. Zhang, J. M. Dai, J. Yang, C. Z. Bi, Y. J. Ge, X. G. Qiu, J. J. Du, and Y. P. Sun, *Phys. Rev. B* 69, 134404 (2004).

[37] J. Deisenhofer, D. Braak, H.-A. Krug von Nidda, J. Hemberger, R. M. Eremina, V. A. Ivanshin, A. M. Balbashov, G. Jug, A. Loidl, T. Kimura, and Y. Tokura, *Phys. Rev. Lett.* 95, 257202 (2005).

[38] Z. G. Sheng, M. Nakamura, F. Kagawa, M. Kawasaki, and Y. Tokura, *Nat. Commun.* 3, 944 (2012).

[39] M. B. Salamon, and M. Jaime, *Rev. Mod. Phys.* 73, 583 (2001).

[40] W. Prellier, Ph. Lecoeur, and B. Mercey, *J. Phys.: Condens. Matter* 13, 915 (2001).

[41] A.-M. Haghiri-Gosnet, and J.-P. Renard, *J. Phys. D: Appl. Phys.* 36, 127 (2003).

[42] M. H. Plan, and S. C. Yu, *J. Magn. Magn. Mater.* 308, 325 (2007).

[43] S. P. Jiang, *J. Mater. Sci.* 43, 6799 (2008).

[44] S. Balevicius, N. Zurauskiene, V. Stankevic, S. Kersulis, V. Plausinaitiene, A. Abrutis, S. Zherlitsyn, T. Herrmannsdorfer, J. Wosnitza, and F. Wolff-Fabris, *Appl. Phys. Lett.* 101, 092407 (2012).

In: Manganite-Based Thin Films ...
Editors: Z. Sheng, L. Hu and Y. Sun
ISBN: 978-1-63482-452-1

Chapter 2

STRAIN INDUCED EFFECTS IN MANGANITE-BASED FILMS AND SUPERLATTICES

***L. Hu*[1] *and Y. P. Sun*[1,2,*]**
[1]Key Laboratory of Materials Physics,
Institute of Solid State Physics, Chinese Academy of Sciences,
Hefei, People's Republic of China
[2]High Magnetic Field Laboratory, Chinese Academy of Sciences,
Hefei, People's Republic of China

Abstract

The presence of strong electron-lattice coupling in perovskite manganites implies the importance of lattice distortions as to the physical properties of these materials. The lattice distortions can be varied by applying hydrostatic or chemical pressure to bulk materials. For thin epitaxial perovskite manganites films, a substrate induced strain is an effective tool to modify the electron-lattice coupling. A wide range and type of strain can be established at discrete and continuous values as a result of the lattice mismatch between the film and the substrate. A brief introduction to the crystal structure of the ABO_3 perovskite is given with special attention to the lattice distortion. And the strain effects on the FMM, COOI, and phase separation states of manganites are also elucidated.

*E-mail address: ypsun@issp.ac.cn

PACS: 75.47.Lx, 75.47.Gk, 73.50.-h, 75.70.-i

Keywords: manganite, thin films, strain effects

1. Crystal Structure of Manganites and Substrates

The crystal structure of the manganite perovskite class possess a fundamental and functional unit: the corner-sharing MnO_6 octahedra with central Mn ions coordinated by six oxygen ligands. The physical properties of the manganites are strongly coupled to the MnO_6 octahedral units, enabled by the highly correlated lattice, spin, charge, and orbital degrees of freedom. The variation in the shape of the octahedral results from the Jahn-Teller effect when the filling of Mn^{3+}ions leads to an orbital degeneracy. This degeneracy can be lifted by an appropriate structural distortion. This typically manifests as an elongation of some Mn-O bonds and a shortening of others. The associated arrangement of the elongations-called a *cooperative Jahn-Teller*, or an *orbital ordering* determines the resulting symmetry of the system. Rotations of the MnO_6 octahedral buckle the octahedral framework along one or more high symmetry axes to accommodate imperfect packing within the cubic unit cell depending on cationic radii. The changes in the unit-cell symmetry resulting from different octahedral rotations have been systematized by Glazer [1] and later expanded by Woodward [2, 3]. Octahedral rotations in the perovskite-type unit cell can be described using Glazer notation that the tilt system is written as $a^{\#}b^{\#}c^{\#}$ in which the Latin letters specify the rotations about three symmetry axes of the pseuocubic unit cell: $[100]_c$, $[010]_c$, and $[001]_c$, and the superscripts indicate whether adjacent octahedral rotate in-phase (+), out-of-phase (-), or not at all (0). Since the octahedral are connected in three dimensions, a rotation or tilt in one direction restricts the allowed tilts and rotations in other directions. In fact only 23 tilt systems belonging to 15 space groups can be obtained. Some common types are shown in Fig. 2.1. The variation of the MnO_6 octahedral framework can alter the translational symmetry and influence the material properties by bending the Mn-O-Mn bond angle away from 180°, which would modify the hybridization/chemical bonding through changes in the overlap of the Mn 3d and O 2p orbitals [4]. The lattice distortions in bulk materials can be controlled by the judicious choice of the A- and B-site cation size mismatch through chemical pressure or application of external hydrostatic pressure. For the thin epitaxial manganites films, a substrate-induced strain is an effective tool to modify the

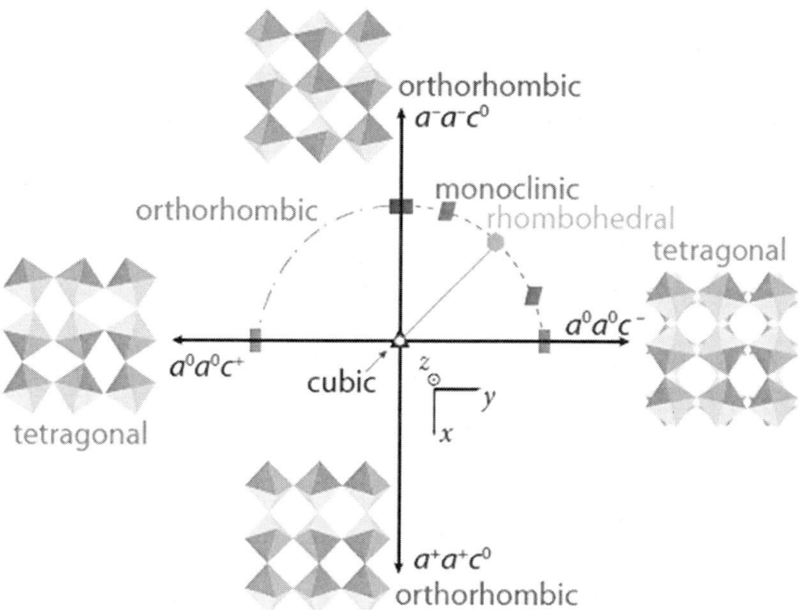

Figure 2.1. Representative octahedral tilt patterns described in the text: $a^0a^0c^+$ and $a^0a^0c^-$ correspond to in-phase and out-of-phase rotations of the octahedra about the z-axis, respectively, while $a^+a^+c^0$ and $a^-a^-c^0$ produce similar rotations of the octahedra in the xy-plane. The relative rotation of adjacent octahedra from one layer to the next is clearly seen for the $a^0a^0c^-$(Reproduced by permission of Reference [4], Copyright 2011, Advanced Materials, WILEY).

lattice distortions and resultant functional behaviors.

Since most applications require thin films with good magnetic and electrical properties, it is of great importance to prepare high quality single-crystal epitaxial films. The growth of high quality manganite thin films has been realized, benefiting from studies of high temperature superconducting thin films. It has been shown that the main physical properties of manganite thin films are often obviously different from their bulk materials. Firstly, the difference is governed by the strain that is due to the lattice mismatch between the film and the single crystal substrate. The strain can strongly influence the crystal structure, Currie temperature (T_C), transport properties, magnetic anisotropy, and phase separation of the manganite thin films. The second difference is the thickness-related finite size effect. The third is the induction of defects (oxygen vacanies) during the deposition of thin films. Therefore, we can tune the properties of manganite thin films via choosing proper substrate, changing the film thickness and oxygen content.

A lot of perovskite single-crystal substrates with structural perfection are commercially available and are well suited for epitaxial growth of manganite thin films. These single-crystal substrates include $YAlO_3$, $LaAlO_3$, $LaSrGaO_4$, $NdGaO_3$, $SrTiO_3$, $DyScO_3$, etc. The lattice parameters of the manganites are very close to those of available substrates. When the manganite thin films are deposited on the single crystal substrates, the in-plane lattice of the thin films would match that of the substrates. The lattice mismatch between the film and the substrate will produce lattice strain. Growth of strained manganite thin film predominantly relies on heteroepitaxy, where a single-crystalline film with specific orientation is deposited on the atomically smooth surface of the substrate. These two materials typically share similar interfacial structure, therefore the structure and crystal orientation of the epilayer are determined by the structure and surface plane of the substrate. By carefully choosing the substrate and growth conditions, coherent strained epitaxial films can be obtained. It is widely accepted that the strain acts by imposing a new in-plane lattice constant on the film, however exactly how that change in lattice constant is exactly accommodated is unclear and difficult to determine experimentally.

The lattice mismatch δ along the interface is defined as $\delta = (a_{substrate} - a_{bulk})/a_{substrate}$. Positive values of mismatch correspond to tensile strains, whereas negative values denote compressive strains. Figure 2.2 (a) and (b) shows that the change in the in-plane lattice parameter is accommodated by a change in the in-plane Mn-O bond lengths. The substrate mismatch strain induces a lattice deformation, which can be analysed from the unit cell distortion. This distortion is usually measured by x-ray diffraction (XRD) studies and high resolution electron microscopy (HREM). The out-of-plane (in-plane) lattice parameter can be precisely determined from the position of the intense [hkl] diffraction peaks in the $\theta - 2\theta$ XRD (in-plane Φ-scans). HREM is used to confirm the parameter values, check the substrate/film interface, and observe dislocations or other defects if present.

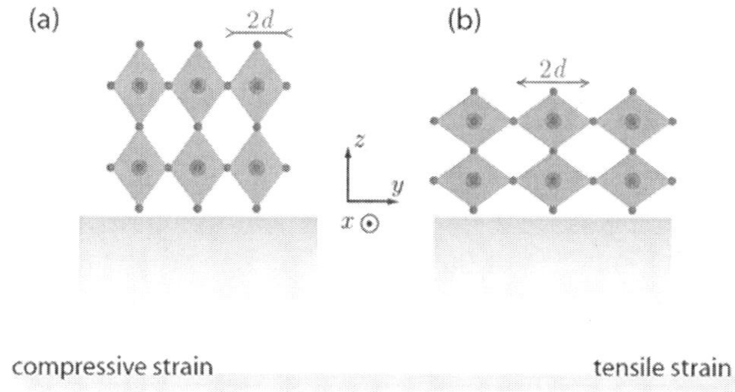

Figure 2.2. Schematic draft of the lattice distortion in a strained epitaxial film under (a) compressive strain or (b) tensile strain(Reproduced by permission of Reference [4], Copyright 2011, Advanced Materials, WILEY).

2. Overview of Growing Methods

The modern study of complex oxide materials has been driven largely by the development of new growth and characterization techniques. The development of new thin film growth techniques that allow for the production of non-equilibrium phases of materials and strain engineering of existing materials have been developed. Numerous film-growth techniques including co-evaporation, sputtering, molecular beam epitaxy (MBE), pulsed-laser deposition (PLD) as well as chemical vapor deposition have been employed in the synthesis of epitaxial manganite thin films. With physical vapor deposition of manganite, the phase constituents are delivered as a flux of individual atoms or simple oxide species. Atomic-level control of the manganite thin film-growth process can be achieved with the help of in-situ growth approaches.

2.1. Pulsed-laser Deposition

PLD is the most versatile, simple, and straightforward approaches for complex oxide film growth. Though significant development of this technique began only in the late 1980s, PLD is now a widely used deposition approach for film deposition. In this technique, an intense pulsed-laser is focused onto a target of the material to be ablated. Above a certain power density, significant material removal occurs in the form of an ejected luminous plume. The threshold power density to produce such a plume depends on the target material, its morphology, and the laser pulse wavelength and duration. It might be of the order of 10-500 $MWcm^{-2}$ for ultraviolet excimer laser pulses of 10 ns duration. The ablated material is ejected from the target in a forward-directed plume. Material from the plume is then to recondense on a substrate, where film growth occurs. The growth process can be supplemented by a passive or reactive gas or ion source, which may affect the ablation plume species in the gas phase or the surface reaction [5]. PLD has several attractive features, including stoichiometric transfer of material from the target of the substrate, generation of energetic species, hyperthermal reaction between the ablated cations and molecular oxygen in the ablation plasma, and compatibility with background pressures ranging from ultrahigh vacuum to 100 Pa [6, 7]. PLD also has technical and fundamental drawbacks: the production of macroscopic ejecta during the ablation process, impurities in the target material, crystallographic defects in the film caused by bombardment by high kinetic energy ablation particles, and inhomogeneous flux and angular energy distributions within the ablation plume.

With the development of new technologies to aid the PLD process, researchers now can create precisely controlled interfaces in material that rival the MBE. One major advance is the incorporation of reflection high energy electron diffraction (RHEED) into the PLD process prompting the term laser-MBE to be coined. The principle of RHEED is that a beam of electrons is accelerated at a glancing angle towards the surface of the substrate. In-situ monitoring the oscillations of the intensity can obtain information about the growth of oxide materials. For the growth of oxide thin film, an oxidizing atmosphere is often required. Firstly, the formation of an oxide thin film need an oxidizing species (ozone, atomic oxygen, molecular oxygen) as a component of the flux. The amount of oxygen required for phase formation depends on the thermodynamic stability of the desired oxide phase. Interaction of ablated species with the background oxygen can produce sub-oxides species, which facilitate oxide phase formation. Secondly, the background gas can be used to reduce the kinetic energies of the ablated species. Time-resolved spectroscopy studies have shown that the background gas can be used to reduce the plume energies from the order of several 100 eV to less than 1 eV.

2.2. Sputtering

The basic sputtering process has been known and used for many years. In sputter deposition, energetic ions created in an rf or dc plasma bombard a metal or oxide target surface. This process ejects atoms from the target that subsequently deposit on a nearby substrate surface. Several sputter deposition techniques have been used in the growth of oxide films including on-axis dc magnetron sputtering, cylindrical magnetron sputtering, ion-beam sputtering and off-axis sputtering. In an on-axis geometry, the substrate and target are facing each other. This is the optimal geometry for maximum deposition rate, but can result in film damage due to the bombardment of the film surface with energetic species from the plasma. An alternative is an off-axis geometry, in which the substrate surface is oriented perpendicular to the surface of the sputter target. This removes the film from the plasma region, eliminates sputter damage, and generally results in films with fewer defects. One disadvantage of sputter deposition is that the stoichiometry of a multi-component film is not necessarily that of the target material due to differences in sputtering yields for different elements. This is often compensated for by using a non-stoichiometric target. Key to the ability to create high quality oxide thin films is intimate knowledge of sputtering yields of various chemical species in a material (as different elements are sputtered at different rates from targets, starting composition may need to be carefully tuned to give stoichiometric final films), deposition rate is crucial for controlling the crystal phase, temperatures for microstructure, sputtering atmosphere is important in controlling surface structure, and substrate position or bias voltage is critical to determining the types and density of defects in these films.

2.3. MBE

MBE generates low energy ($\sim$1 eV) thermal atomic beam for each of the constituent materials (both cations such as transition metals and anions such as oxygen) desired in the film. Beams of metal atoms can be created using radiatively heated sources (i.e., Knudsen or effusion cells) or from electron beam evaporators. Oxygen is typically supplied as molecular oxygen (O_2), ozone (O_3), or atomic oxygen (O) due to greatly enhanced oxidizing potentials ozone and atomic oxygen are commonly used today to assure the growth of a fully oxidized film. This technique offers atomic-level control of the film-growth process and can be used to produce extremely pure and defect-free film.

2.4. Metal-organic Chemical Vapor Deposition

For large-scale production of thin films, metal-organic chemical vapor deposition (MOCVD) is very attractive. It is routinely utilized in the electronics industry and is quite amenable to large area deposition with high throughput. MOCVD is independent of line-of-site deposition and can be used for in situ growth at oxygen pressures near 1 atm. With MOCVD, the cations necessary for film growth are delivered as constituents of organometallic molecules. If the organometallic molecules are sufficiently volatile, they can be delivered to the heated substrate via a carrier gas. For non-volatile precursors, the reactants are delivered as a condensed phase. The molecules thermally decompose at the heated substrate surface resulting in film growth. For oxide film growth, an oxidizing species is included within the gas flow. MOCVD has the advantage over physical vapor deposition techniques in being conformal. This yields a uniform deposition rate over all exposed surfaces of a non-planar structure. One significant challenge in the synthesis of oxide films using MOCVD is the development and reproducible synthesis of volatile precursor molecules that are stable in storage and transport and decompose at elevated temperatures to yield good films with no contamination from the organic ligand.

3. Strain Effects in Manganites

Depending on the A-site ionic radii and hole-doping level, the manganites exhibit various kinds of electronic phases such as ferromagnetic metal (FMM), paramagnetic insulator (PMI), charge- and orbital-ordered insulator (COOI), and ferromagnetic insulator (FMI). These competing phases originate from the strong coupling among charge, spin, orbital and lattice degrees of freedom. Many experimental and theoretical studies have shown that the substrate-induced strain strongly affects the physical properties of manganite thin films. Generally, the method for the experimental study of biaxial strain effects is the growth of epitaxial thin films on single-crystalline substrates with a certain lattice mismatch, with the thickness of these films vary over a relatively large range. Let us now focus on the influence of strain on both the electrical and magnetic properties of manganite thin films.

3.1. Strain Effects on the FMM States of Manganites

The CMR manganites $R_{1-x}A_xMnO_3$, with R a rare earth such as La or Nd and A a divalent alkali such as Sr or Ca, are of particular current interest because in the regime $0.2 < x < 0.5$ they are fully spin-polarized FMM at low temperature and PMI at high temperature. In the vicinity of the Curie temperature T_C, a concurrent metal-insulator transition (MIT) occurs and the MR is extremely large [8]. Millis *et. al* have developed a model to predict the dependence of a biaxial strain on T_C [9]. The three-dimensional strain states can be decomposed into a bulk strain $\epsilon_B = \frac{1}{3}(2\epsilon_{xx} + \epsilon_{zz})$ and a biaxial distortion $\epsilon^* = \frac{1}{2}(\epsilon_{zz} + \epsilon_{xx})$. The effects of strain on the T_C can be described by the formula

$$T_C(\epsilon_B, \epsilon^*) = T_C(0,0)[1 - \alpha\epsilon_B - b\epsilon^{*2}] \qquad (1)$$

with $\alpha = (1/T_C)[dT_C/d\epsilon]$ and $b = (1/T_C)[d^2T_C/d\epsilon^{*2}]$. A hydrostatic compression tends to increase the hopping and thereby reduce the electron-lattice coupling inducing an increase of both the bandwidth and T_C. Therefore, depending on the sign of the strain the T_C change associated with ϵ_B will be positive or negative, respectively. On the other hand, a biaxial distortion ϵ^* will increase the Jahn-Teller splitting of the e_g electron levels which reinforce the electrons tendency to become more localized decreasing T_C. Typical values of α and b in manganites were predicted to be about 6 and 1.4×10^3, respectively.

Adamo *et al.* have deposited 22 nm $La_{0.7}Sr_{0.3}MnO_3$ thin films on different single crystalline substrates, varying the substrate-induced biaxial strain from -2.3% to +3.2%. The ratio a_{zz}/a_{xx} between the out-of-plane and in-plane film lattice parameters as a function of the substrate lattice parameter is shown in Fig. 2.3(a). The regular behavior of the tetragonal distortion is a clear indication of the dominant role played by biaxial strain on the structural lattice deformations of the films. This is further confirmed by the results shown in Fig. 2.3(b). The influence of biaxial strain on the electrical transport and magnetic properties of the thin films grown on various substrates is shown in Fig. 2.4. The results clearly indicate that the strain has a significant effect on T_P and T_C. In the case of thin films with a small amount of strain ($SrTiO_3$, LSAT, and $NdGaO_3$), the low temperature resistivity values are comparable to those measured on single crystals. The T_P is higher than 390 K for the samples grown on $NdGaO_3$ and LSAT. T_P and T_C are reduced as the magnitude of the substrate mismatch increases. In Fig. 2.3(c), the plane in the figure is obtained by fitting the experimental data using Eq. (1). From the fitting procedure, the

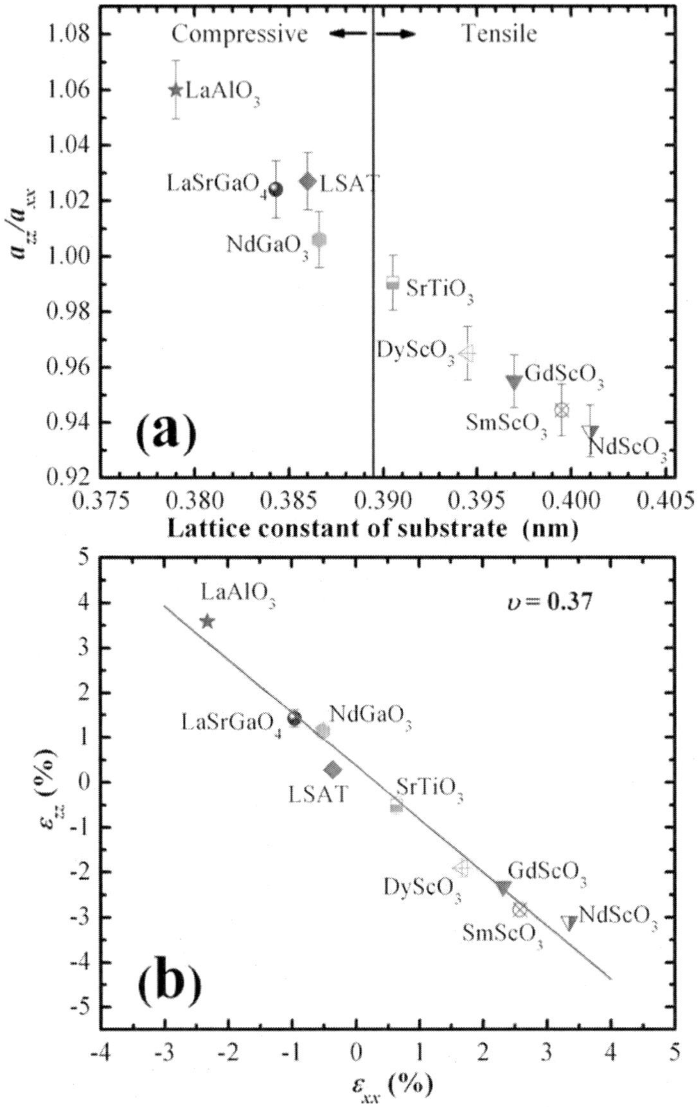

Figure 2.3. (a) Ratio between the out-of-plane (a_{zz}) and in-plane (a_{xx}) lattice spacing of commensurately $La_{0.7}Sr_{0.3}MnO_3$ films as a function of the pseudocubic in-plane substrate lattice parameter. (b) The out-of-plane lattice strain ϵ_{zz} as a function of the in-plane lattice strain ϵ_{xx}(Reproduced with permission from Reference [10]. Copyright 2009, AIP Publishing LLC).

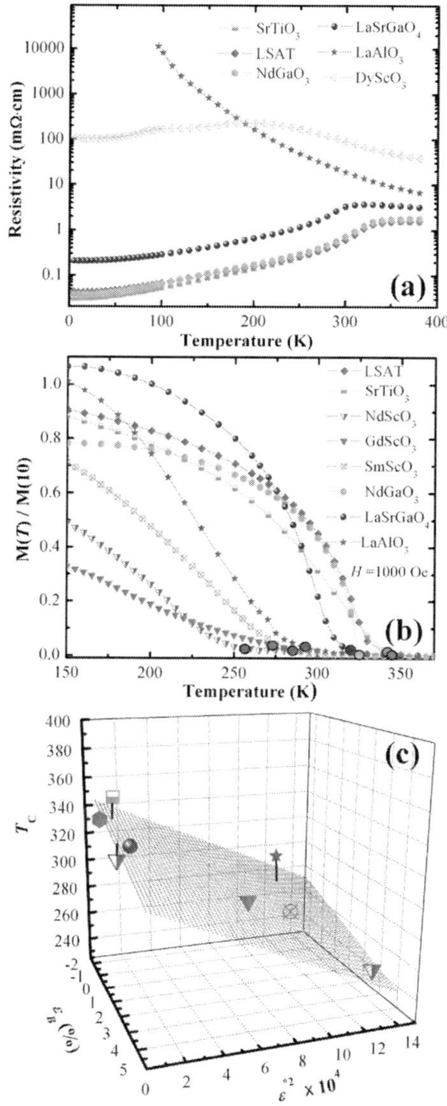

Figure 2.4. (a) Resistivity vs temperature behavior, at zero applied magnetic field, of 22 nm thick $La_{0.7}Sr_{0.3}MnO_3$ films on different substrates. (b) The temperature dependence of the magnetization normalized at 10 K of samples cooled in a 1000 Oe field of H=1000 Oe. Circles indicate the Curie temperature, T_C. (c) The Curie temperature T_C vs the ϵ_B and ϵ^* strains. The best fit plane to the data is also shown(Reproduced with permission from Reference [10]. Copyright 2009, AIP Publishing LLC).

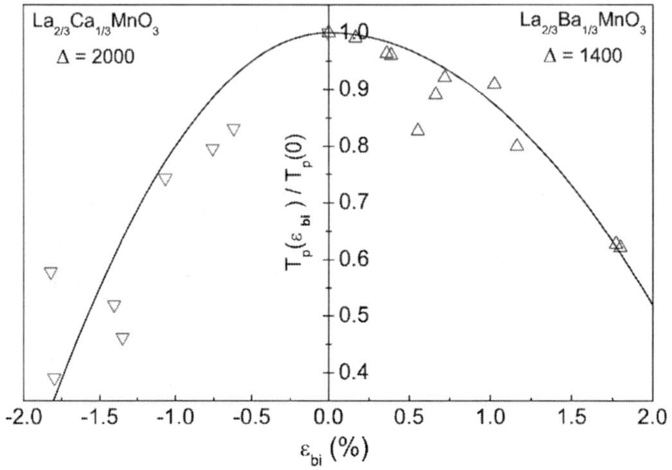

Figure 2.5. MIT temperature T_P vs biaxial distortion(Reprinted with permission from Reference [16], Copyright (2006) by the American Physical Society).

obtained $T_C(0,0) = 345 \pm 9$ K, $\alpha = 1.55 \pm 0.01$, and $b = 1460 \pm 30$,are in very good agreement with the predictions of Millis [10]. The coherently strained state can be kept only up to a critical thickness h_c. Increasing the film thickness above h_c results in the strain relaxation accompanied by the formation of misfit dislocations and structural disorder [11, 12, 13, 14, 15]. To overcome this problem, Lu *et al.* have grown a series of high-quality, coherent strain [$La_{2/3}$(Ca or Ba)$_{1/3}$$MnO_3$/$SrTiO_3$]×N superlattices on (100) $SrTiO_3$ and $NdGaO_3$ substrates by laser MBE. The biaxial strain (tensile and compressive) can be varied over a wide range by varying the thickness ratio of the constituent layers. It is found that the MIT temperature T_P decreases considerably with increasing biaxial distortion as $T_P/T_P(0) = 1 - b\frac{1}{2}\epsilon^{*2}$ for both tensile and compressive strain with b of the order of 1500, as shown in Fig. 2.5 [16].

The above methods to vary the strain states in manganite thin films rely on different types of substrates or different film thickness. The defect of the crystal lattice is a second variable parameter, which is likely to affect magnetic and electric properties and will hinder the quantification of the strain effects. An alternative strategy is to deposit manganite thin films on piezoelectric substrate, such as $BaTiO_3$ and Pb($Mg_{1/3}Nb_{2/3}$)O_3-$PbTiO_3$ (PMN-PT), and employ its structural transitions and piezoelectric effect to change the strain states

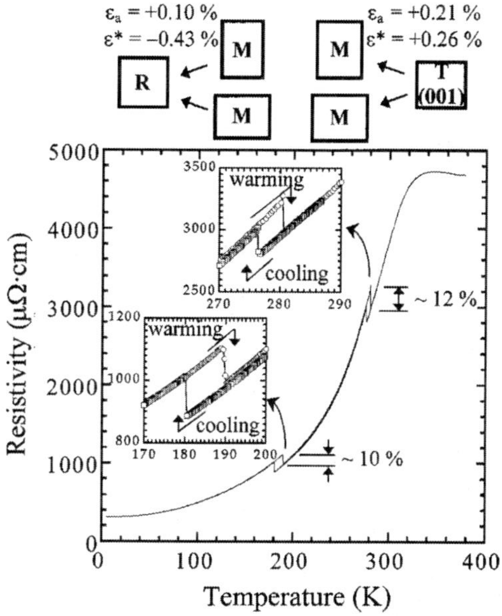

Figure 2.6. Electrical resistivity as a function of temperature of a 50 nm $La_{0.67}Sr_{0.33}MnO_3$ film grown on (001) $BaTiO_3$ substrate. The insets are the zoomed-in view of the thermal hysteresis at the two transitions, and the corresponding in-plane lattice distortions of (001) $BaTiO_3$ are schematically depicted above the graph. The resistivity data were obtained from six thermal cycles(Reproduced with permission from Reference [17]. Copyright 2000, AIP Publishing LLC).

[17, 18, 19, 20, 21, 22, 23, 24, 25]. This approach makes it possible that the intrinsic effects of the substrate-induced strain on the properties of films can be studied without introducing the effects of extrinsic variables, e.g., crystalline quality, disorder, and oxygen nonstoichiometry. Lee *et al.* have studied the effects of induced biaxial strain on the electrical transport and magnetic properties of epitaxial thin film of $La_{0.67}Sr_{0.33}MnO_3$ by structural transitions of ferroelectric $BaTiO_3$ substrates. Large jumps of electrical resistivity ($\sim$12%) and low field magnetization ($\sim$70%) have been observed in the films at the structural transition temperatures of $BaTiO_3$ substrate (Fig. 2.6) [17]. Thiele *et al.* have found that the epitaxially grown $La_{0.7}Sr_{0.3}MnO_3$ thin films show resistance modulations induced by the inverse piezoeffect of the employed PMN-PT substrates (Fig. 2.7). The in-plane strain state of the films can be continuously

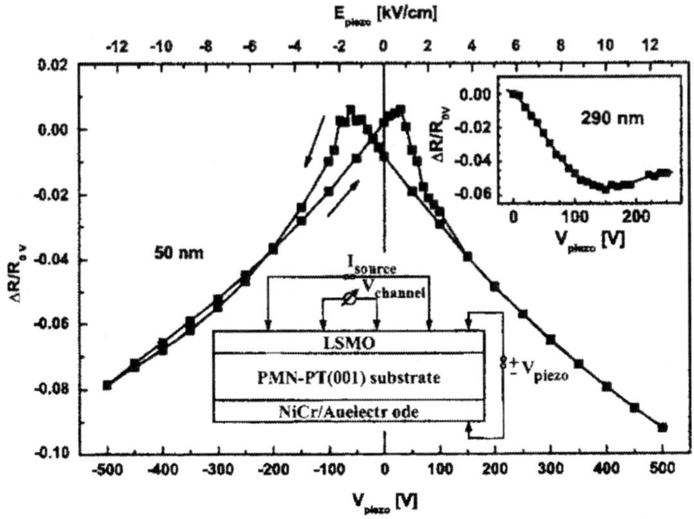

Figure 2.7. Resistance loop in dependence on applied piezovoltage for a 50 nm thick $La_{0.7}Sr_{0.3}MnO_3$ film on PMN-PT(001). The film is under tensile strain at V_{piezo}=0 and relaxes gradually with increasing piezovoltage(Reproduced with permission from Reference [19]. Copyright 2005, AIP Publishing LLC).

tuned by application of a piezovoltage to PMN-PT. Variation of in-plane lattice parameters by $\sim$0.06% reversibly changes the resistance of the films by up to 9% at 300 K and shifts the magnetic T_C [19]. Zheng *et al.* have used ferroelectric PMN-PT single crystals as substrates and studied the effects of substrate-induced strain on the strain state, resistance, T_C, and magnetotransport properties of $La_{0.7}Ba_{0.3}MnO_3$ films by *in-situ* induced strain in the film via the ferroelectric poling or the converse piezoelectric effect . The experimental results indicate that, associated with reduction in the in-plane tensile strain, T_C increases linearly with increasing E ($\delta\epsilon_{xx}$) as shown in Fig. 2.8. Since the reduction in the in-plane tensile strain $\delta\epsilon_{xx}$ is proportional to E, the relationship between T_C and $\delta\epsilon_{xx}$ can be expressed as $\Delta T_C/T_C$=q$\delta\epsilon_{xx}$, where q is found to be 11.4. This indicates that a change in the in-plane strain by 1% would induce a 11.4% shift in T_C, which is consistent with the theoretical prediction by Millis *et al.* [9] who showed that a change in the in-plane biaxial strain by 1% would cause a 10% shift in T_C, and thus provides evidence for the important role of the Jahn-Teller electron-lattice coupling linked to the substrate-induced strain in the LBMO film [23].

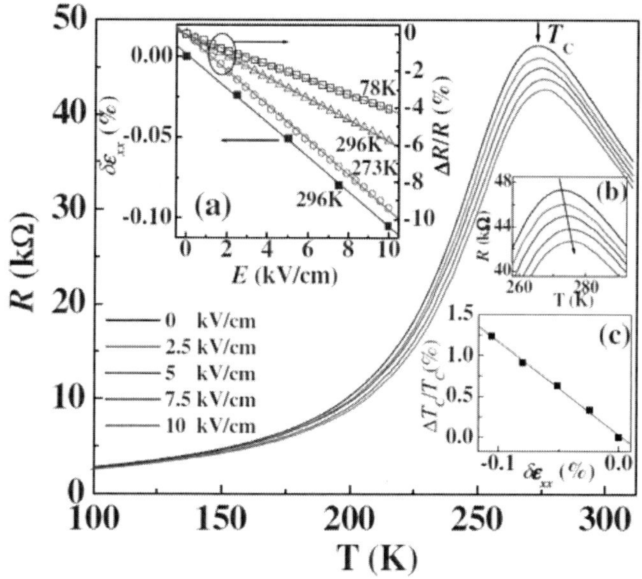

Figure 2.8. Temperature dependence of the resistance for the LBMO film when a dc electric field E is applied to the LBMO/PMN-PT structure after the PMN-PT substrate had been polarized to P_r^+. The inset (a) shows the relative change of the resistance $\Delta R/R$ and the induced reduction in the in-plane tensile strain, $\delta\epsilon_{xx}$, of the LBMO film versus E. The inset (b) shows an expanded view of the temperature dependence of the resistance near T_C. The inset (c) shows $\Delta T_C/T_C$ as a function of $\delta\epsilon_{xx}$(Reprinted with permission from Reference [23], Copyright (2009) by the American Physical Society).

Besides the field-induced strain effect, the polarization effect might also be considered as a result of using ferroelectric crystal. Sheng *et al.* have showed that the coaction of electric field induced strain and polarization effects in $La_{0.7}Ca_{0.3}MnO_3$/PMN-PT structures. The strain state of the films can continuously be tuned by the application of electric field. Figure 2.9 shows the results of transport measurements of LCMO film under bias electric field. The resistance R at room temperature drops by about -3.61% at E=-8 kV/cm due to the electric-induced in-plane strain in the film. A butterflylike hysteresis loop with coercive fields of 2 kV/cm is also observed when cycling the electric field

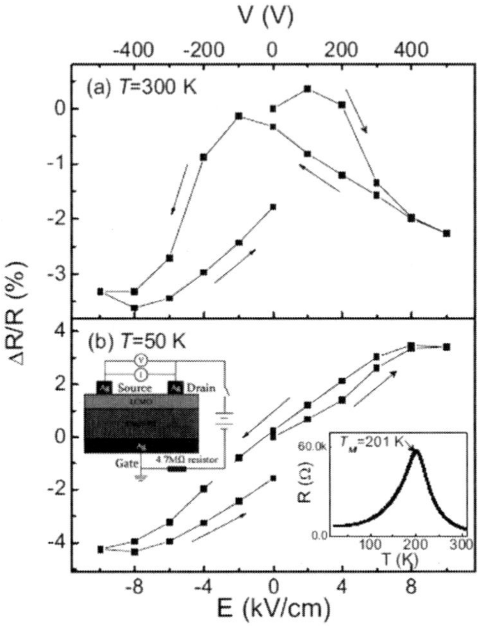

Figure 2.9. Resistance hysteresis of the LCMO film vs bias electric field at (a) 300 K and (b) 50 K. The left inset of (b) is the schematic for the LCMO-PMNT structure in FET configuration and the circuit for electrical measurements. The right inset of (b) shows the typical temperature dependence of electrical resistance R-T of LCMO film(Reprinted with permission from Reference [24], Copyright (2009) by the American Physical Society).

from the positive to negative values. As temperature decreases, the resistance of LCMO film decreases (increasing) with increasing negative (positive) bias voltage to exhibit shuttlelike loop. Figure 2.10 shows the response of magnetization of LCMO/PMN-PT structure to the external electric field at T=10 K. When a E= +9.6 V/cm bias is applied, a sizeable decrease in the magnetization is observed, and the magnetization shows a quick switch between the original state and lower magnetization state corresponding to the bias on and bias off. The lower magnetization state is stable for a constant bias voltage, unaffected by the repeated bias voltage on and off. The relative variation $\Delta M/M(0)$ is about -8.64%. When a negative bias electric field E=-9.6 kV/cm is applied, the relative variation is about -7.47%. The results indicate that the electric field induced strain effect dominates the magnetization when the LCMO is in FMM

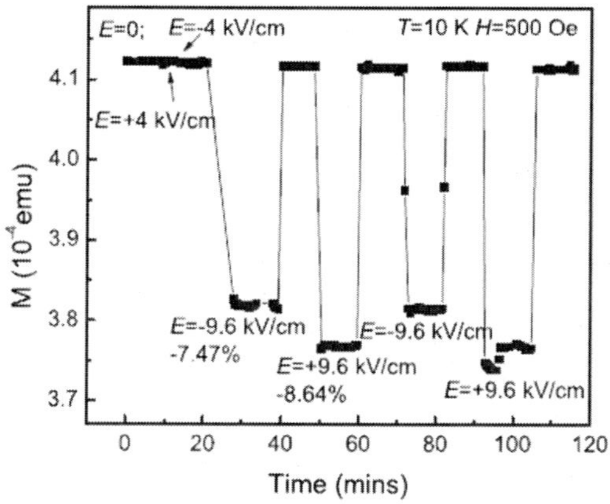

Figure 2.10. Magnetization of the LCMO/PMN-PT structure as a function of time measured at 10 K with a magnetic field of 500 Oe(Reprinted with permission from Reference [24], Copyright (2009) by the American Physical Society).

state, while the polarization effect dominates for the electrical transport properties [24].

The substrate-induced strain can not only modulate the T_P and T_C, but also create phases not observed in bulk materials [26, 27, 28, 29, 30, 31, 32, 34]. Konishi *et al.* have showed that magnetic and electronic phases can be controlled by the epitaxial strain for thin films of the manganite, $La_{1-x}Sr_xMnO_3$, grown on perovskite substrates with various lattice parameters. The strain-induced orbital-ordering (disordering) via coupling with the cooperative Jahn-Teller distortion produces different spin-structures, such as ferromagnetic, layer-type or chain-type antiferromagnetic states, and hence different magnetoelectronic characteristics. The spin-orbital phase diagram for $La_{1-x}Sr_xMnO_3$ films has been obtained in the plane of strainfield (c/a ratio) vs doping 4 x, which can be quantitatively reproduced by the density-functional electronic-calculation [26, 27] (Fig. 2.11). The saturation magnetization

and conductivity of $La_{0.7}Sr_{0.3}MnO_3$ films decrease with decreasing thickness. Huijben *et al.* have precisely determined the critical thickness for a nonmetallic nonferromagnetic layer (dead layer) at the interface with the $SrTiO_3$ (001) substrate of only three unit cells through detailed analysis of transport, magnetism, and x-ray absorption spectroscopy measurements on ultrathin $La_{0.7}Sr_{0.3}MnO_3$ films with thicknesses from 3 to 70 unit cells (Fig. 2.12) [30]. Such behaviors are attributed to the orbital ordering or orbital re-

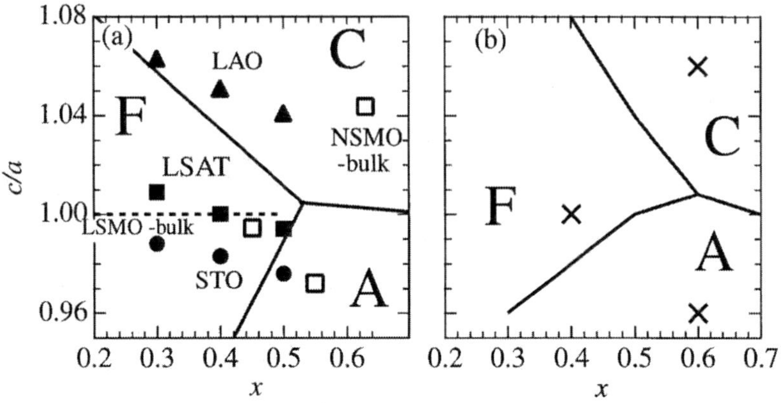

Figure 2.11. (a)The extended phase diagram of $La_{1-x}Sr_xMnO_3$ (closed symbols) in the plane of the anisotropic lattice strain (c/a ratio) vs doping level (x). Open squares and a broken line represent the results for the bulk single-crystals $Nd_{1-x}Sr_xMnO_3$ (NSMO) and $La_{1-x}Sr_xMnO_3$ (LSMO). (b) The corresponding calculated magnetic phase diagram based on the density functional method assuming the experimental volume (Reprinted with permission from Reference [26], Copyright (1999) by the Physical Society of Japan).

construction based on the detailed analysis of liner dichroism by x-ray absorption spectroscopy in the ultrathin LSMO films [30, 31]. The dead layer of $La_{0.67}Sr_{0.33}MnO_3$ grown on $LaAlO_3$ substrate can be modulated by magnetic filed and electric voltage [29]. As a result, though $La_{1-x}Sr_xMnO_3$ film is structurally a single-layered film, its electronic distribution is not uniform due to the formation of an intermediate layer of about a few unit cells. Lee *et al.* present an unreported magnetic configuration in epitaxial $La_{1-x}Sr_xMnO_3$ (x∼0.3) (LSMO) films grown on strontium titanate (STO). X-ray magnetic circular dichroism indicates that the remanent magnetic state of thick LSMO films is opposite to the direction of the applied magnetic field. Spectroscopic and scattering measurements reveal that the average Mn valence varies from mixed Mn^{3+}/Mn^{4+} to an enriched Mn^{3+} region near the STO interface, resulting in a compressive lattice along the a, b axis and a possible electronic reconstruction in the Mn e_g orbital ($d_{3z^2-r^2}$) [32]. Furthermore, Boschker *et al.* present a study of the thickness dependence of magnetism and electrical conductivity in ultrathin $La_{0.67}Sr_{0.33}MnO_3$ films grown on $SrTiO_3$ (110) substrates. It is found that a critical thickness of 10 unit cells below which the conductivity of

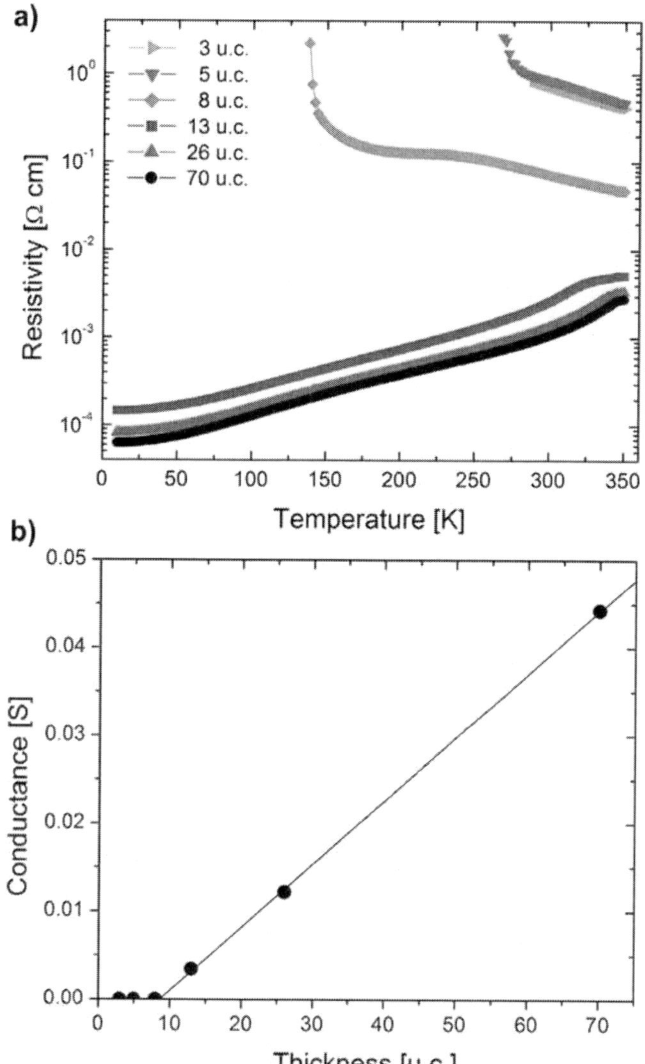

Figure 2.12. Transport properties of ultrathin LSMO films on $SrTiO_3$ (001) substrates. (a) Temperature-dependent resistivity for films of different thicknesses. (b) Thickness dependence of the total conductance of films at 10 K (Reprinted with permission from Reference [30], Copyright (2008) by the American Physical Society).

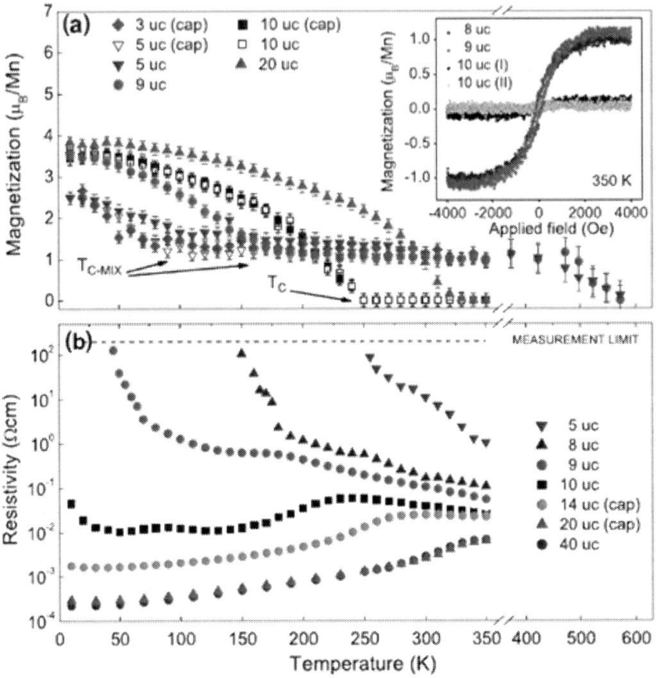

Figure 2.13. Temperature dependent magnetization and resistivity measurements of LSMO (110) thin films. (a) Saturation magnetization of various samples with different thicknesses. The inset shows magnetic hysteresis loops of the n=8, 9, and 10 uncapped samples at T=350 K. (b) Resistivity of various samples with different thicknesses (Reprinted with permission from Reference [34], Copyright (2012) by the American Physical Society).

the films disappeared and simultaneously the Curie temperature increased, indicates a magnetic insulating phase at room temperature. These samples have a Curie temperature of about 560 K with a significant saturation magnetization of $1.2 \pm 0.2 \mu_B$ (Fig. 2.13). The canted antiferromagnetic insulating phase in ultra thin films of n<10 coincides with the occurrence of a higher symmetry structural phase with a different oxygen octahedra rotation pattern[34]. In addition to the static structural change of film due to the lattice mismatch, the dynamic lattice vibration (phonon) of the substrate can extend across an epitaxial interface into the manganite thin film to affect its physical properties. Segal *et al.* have studied the transport and magnetic properties of $La_{0.53}Sr_{0.47}MnO_3$

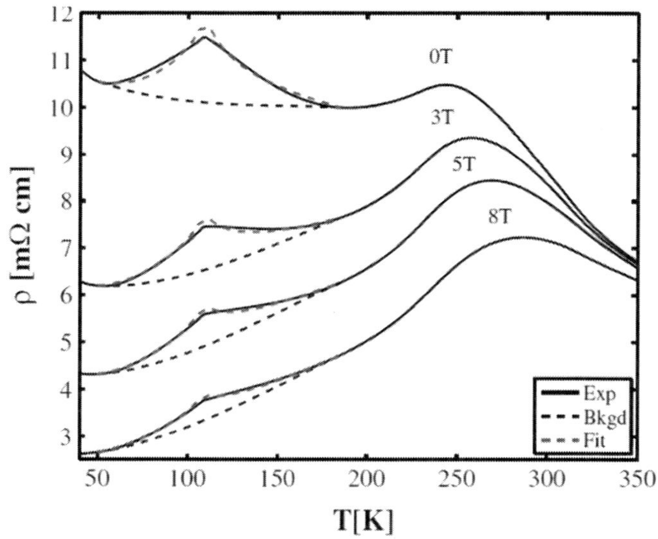

Figure 2.14. Resistivity of an 11 uc film of $La_{0.53}Sr_{0.47}MnO_3$ as a function of temperature, for several out-of- plane magnetic field values. The PZT overlayer is in the depletion state. Black dashed lines are interpolations excluding the cusp, and the gray (red) dashed lines are fitted to A_{25} phonon coupling(Reprinted with permission from Reference [33], Copyright (2011) by the American Physical Society).

ultrathin films grown epitaxially on $SrTiO_3$ and found that a sharp cusp at the structural transition temperature of the substrate (Fig. 2.14). Using a combination of experiment and first principles theory they show that the cusp is a result of evanescent cross-interface coupling between the charge carriers in the film and a soft phonon mode in the $SrTiO_3$, which is mediated through linked oxygen octahedral motions[33].

3.2. Strain Effects on the COOI Phases of Manganites

The COOI phase in manganites, a long-range ordering of the Mn^{3+} and Mn^{4+} ions and d-electron orbits, is the result of a complicated competition between Coulomb interactions (between the charges), exchange interactions (between the Mn moments) and the electron-lattice coupling through Jahn-Teller distortions of the oxygen octahedron surrounding the Mn^{3+}. Generally, the COOI

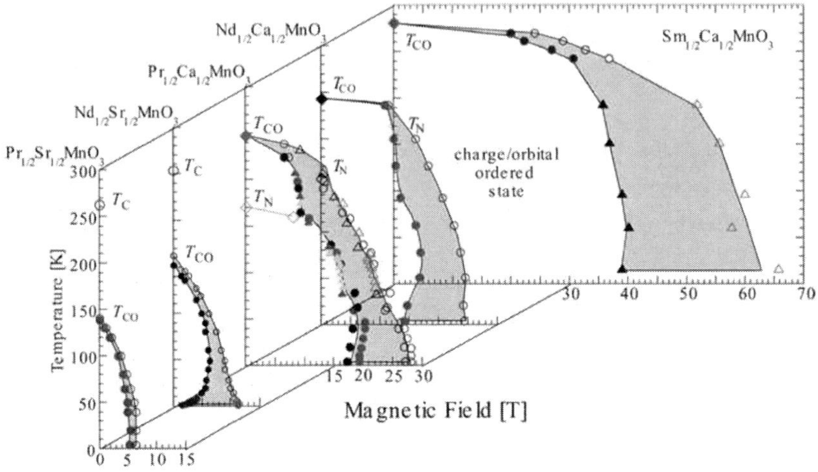

Figure 2.15. The COOI phase of various $R_{1/2}A_{1/2}MnO_3$ plotted on the magnetic field-temperature plane (Reprinted from Reference [35], Copyright (1999), with permission from Elsevier).

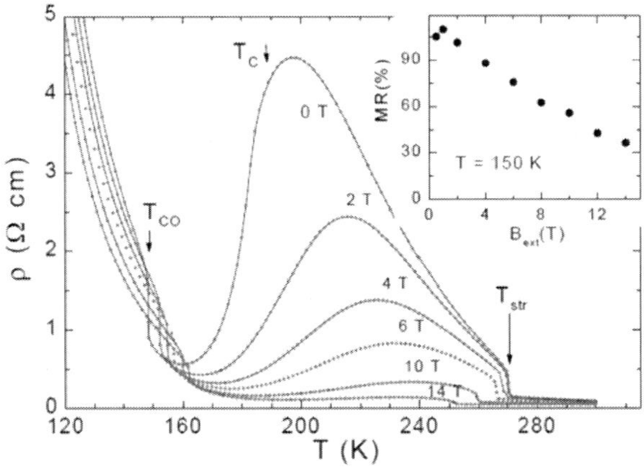

Figure 2.16. Resistivity versus temperature for $La_{0.875}Sr_{0.125}MnO_3$. The transition temperatures T_{str} for the structural, T_C for the ferromagnetic, and T_{CO} for the charge ordering transition are marked by the arrows. The inset shows the magnetic field dependence of the positive MR at T_{CO} (Reprinted with permission from Reference [41], Copyright (1999) by the American Physical Society).

phase appears mostly in narrow and intermediate bandwidth manganites, e.g., $Pr_{1-x}Ca_xMnO_3$. It often competes with the (FM) double exchange (DE) interaction and gives rise to an insulating ground state with antiferromagnetism. This kind of COOI phase can be melted by an external magnetic field to result in an insulator-metal transition (Fig. 2.15) [35, 36]. Otherwise, there exists another type of COO phase which appears in wide bandwidth manganites, for instance, lightly doped $La_{1-x}Sr_xMnO_3$ (Fig. 2.16). This COOI phase can coexist with the isotropic three-dimensional FM state in spite of the insulating behaviour. Remarkably, this kind of COOI phase can be stabilized by an applied magnetic field, which is distinct from the former type in $Pr_{1-x}Ca_xMnO_3$ [38, 39, 40, 41, 42].

Since the appearance of COOI phase is always accompanied by changes in lattice parameters, it is expected that the strain can significantly influence the COOI phase. Prellier *et al.* have deposited $Pr_{0.5}Ca_{0.5}$MnO3 onto (100)-oriented $SrTiO_3$ substrates. Magnetization and transport properties are measured when the thickness of the film is varied. The thinner films do not exhibit any temperature-induced insulator-metal transition under an applied magnetic field up to 9 T. For thickness larger than 1100 Å, 5 T magnetic field is sufficient to melt the COOI state. It indicates that the robustness of the COOI state in thin films depends strongly on the strains and the thickness (Fig. 2.17) [43]. Yang *et al.* have studied the effect of biaxial strain on the charge ordering temperature T_{CO} of $Pr_{0.5}Ca_{0.5}MnO_3$ thin films grown on $SrTiO_3$. Transport measurements combined with data from electron microscopy show that T_{CO} is considerably enhanced. At a thickness of the order of 10 nm, where the films are fully strained, T_{CO} is above 320 K, more than 70 K above the bulk value of 250 K. While around 50 nm, where relaxation has been set in, the enhancement is around 40 K. The bulk value is only reached at a thickness of about 150 nm (Fig. 2.18) [47]. Okuyama *et al.* have studied the effect of growth orientation on the COOI phase for $Pr_{0.5}Ca_{0.5}MnO_3$ epitaxial thin films fabricated on LSAT substrates by means of resistivity, synchrotron XRD, and polarized optical microscopy measurements. COO transition is observed around 220 K for a film grown on LSAT (011) substrate, similar to a bulk sample, while a film grown on a (001) plane of LSAT shows much higher transition temperature around 300 K. The domain size of COO is approximately three times as large in the (011) film as in the (001) film. These results demonstrate that various properties of COOI phenomena can be controlled with the growth orientation via the epitaxial strain from the substrate (Fig. 2.19)[50].

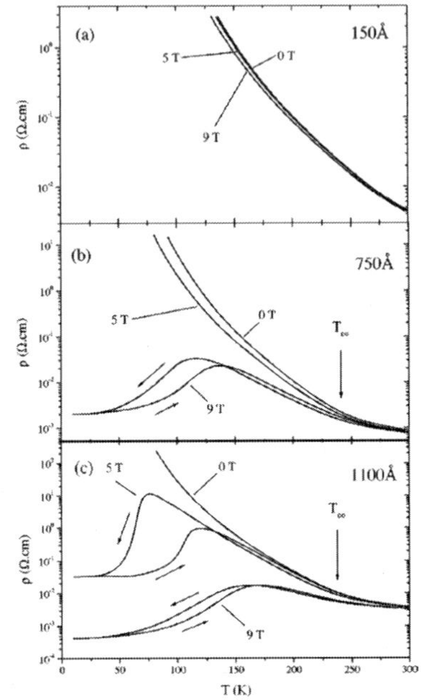

Figure 2.17. ρ(T) under different magnetic field (0, 5, and 9 T) applied in the plane of the substrate for different films (a) 150 Å, (b) 750 Å, and (c) 1100 Å. Arrows indicate the direction of the temperature variation(Reprinted with permission from Reference [43], Copyright (2000) by the American Physical Society).

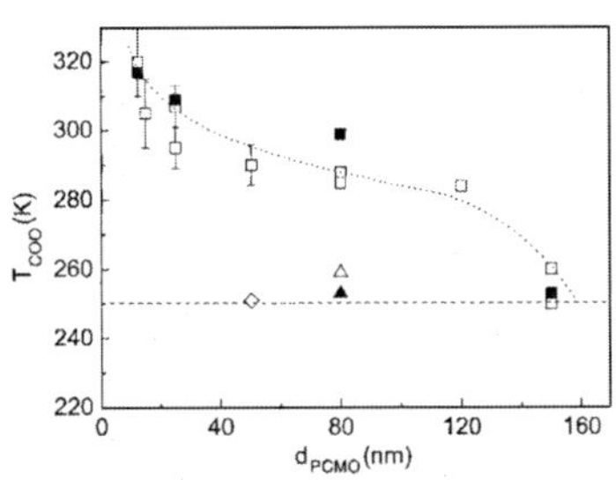

Figure 2.18. Collection of values for T_{COO} for films of PCMO with different thickness d_{PCMO} on different substrates. Open symbols are taken from resistance data, closed symbols from electron microscopy; (□) films on STO; (△) film relaxed by postanneal; (◇) film on $SrLaGaO_4$. The dotted line is a guide to the eye, the dashed line denoted the bulk value for T_{COO}(Reproduced with permission from Reference [47]. Copyright 2006, AIP Publishing LLC).

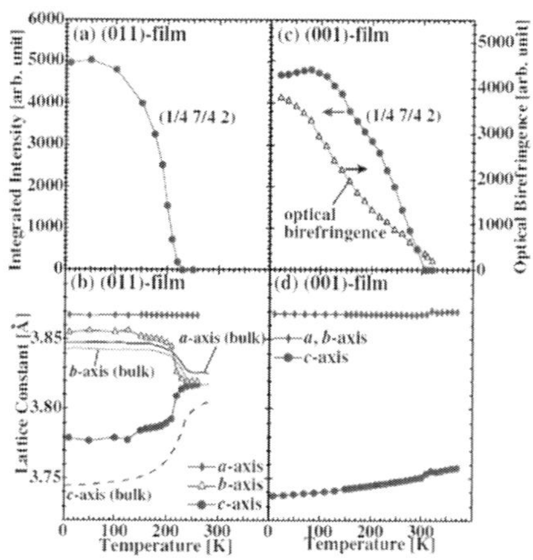

Figure 2.19. Temperature dependence of integrated intensity of ($\frac{1}{4}$, $\frac{7}{4}$, 2) superlattice reflection, which corresponds to the orbital-ordering, for (a) the (011) film and (c) the (001) film. Lattice constants are plotted against temperature for (b) the (011) film and (d) the (001) film(Reproduced with permission from Reference [50]. Copyright 2006, AIP Publishing LLC).

The orbital order (OO) couples intimately to the lattice distortion. It is expected that a tetragonal lattice distortion promotes ferro-orbital structures-the compressive strain within favors $d_{3z^2-r^2}$ (C-OO), while the tensile strain favors $d_{x^2-y^2}$ (A-OO). This case was achieved by manipulating the tetragonal lattice parameters empolying a thin film fabricated on (001) substrates [26, 34]. However, the tetragonal lattice distortion has severe restriction on the lattice deformation, in which only the out-of-plane lattice constant can vary while the in-plane lattice constants are clamped to the substrates. This makes it quite difficult for the films to undergo the transition into the COOI state because a large anisotropic deformation is essential at the transition. The recovery of COOI phase can be realized in intermediate bandwidth $(Nd,Pr)_{0.5}Sr_{0.5}MnO_3$ thin film grown on (011) substrates. Wakabayashi *et al.* have showed that a novel structure of orbital ordering is found in a $Nd_{0.5}Sr_{0.5}MnO_3$ thin film, which exhibits a clear first-order transition, by synchrotron XRD measurements. Lattice parameters vary drastically at the metal-insulator transition at 170 K, and superlattice reflections appear below 140 K (T_{CO}). The accommodation of the large lattice distortion at the first-order phase transition and the appearance of the novel orbital ordering are brought about by the anisotropy in the substrate (Fig. 2.20)[46].

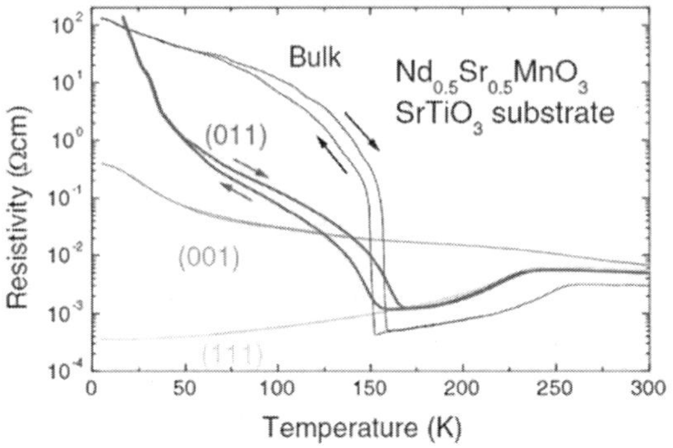

Figure 2.20. Temperature dependence of the electric resistivity of $Nd_{0.5}Sr_{0.5}MnO_3$ thin films grown on $SrTiO_3$ (001), (011), and (111) substrates(Reprinted with permission from Reference [46], Copyright (2006) by the American Physical Society).

For $La_{1-x}Sr_xMnO_3$ compounds, the COOI phase is restricted to a small doping region (0.1⩽x⩽0.15), which is close to a commensurate fraction of charge carriers. It was reported that the formation of COOI phase in lightly doped $La_{1-x}Sr_xMnO_3$ is closely associated with lattice deformation by structural phase transition [37, 38, 39, 40, 41, 42]. For example, neutron diffraction studies reveals that $La_{7/8}Sr_{1/8}MnO_3$ exhibits successive structural phase transitions: high-temperature pseudocubic phase (O*:a∼b∼c/$\sqrt{2}$) to intermediate Jahn-Teller distorted orthorhombic (O*: $b > a$≫c/$\sqrt{2}$) at T_H≈260 K and to low-temperature O* phase at T_{COOI}≈160 K [37].When it was formed as thin films, the COOI transition and lattice deformation would be affected by the tetragonal lattice stability from substrates [48, 55, 56]. Chen *et al.* have realized the COOI phase in $La_{7/8}Sr_{1/8}MnO_3$ thin films by introducing anisotropic strains with (011) $SrTiO_3$ substrates (Fig. 2.21) [48]. Wang *et al.* have investigated the substrate-induced strain effect in $La_{7/8}Sr_{1/8}MnO_3$ thin films grown on (001)-oriented $SrTiO_3$ and PMN-PT substrates. A metal-insulator transition was observed at low temperature in LSMO/PMN-PT, which was ascribed to COOI formation due to a large tensile strain (Fig. 2.22) [55]. Hu *et al.* have studied the effect of compressive strain on COO transition in epitaxial $La_{7/8}Sr_{1/8}MnO_3$ thin films. The compressive strain of LSMO thin films grown on $LaAlO_3$ substrates can be modified by varying the film thickness. It was found that both the COO transition and magnetic transition of LSMO thin films

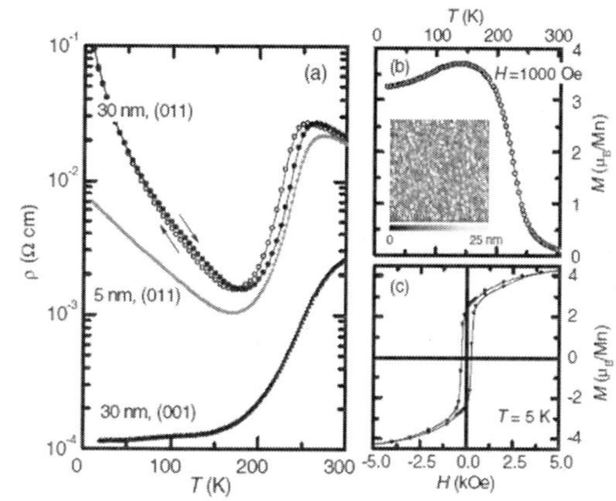

Figure 2.21. The temperature-dependent resistivity of $La_{7/8}Sr_{1/8}MnO_3$ films grown on (001)- and (011)-oriented $SrTiO_3$ substrates; [(b) and (c) the magnetization-temperature curve and the magnetization hysteresis loop for the 30 nm (011) film. The inset of (b) is a 2×2 μm^2 AFM image of the (011) film with a mean-square-root roughness of ~3.8 nm(Reproduced with permission from Reference [48]. Copyright 2008, AIP Publishing LLC).

could be suppressed by the compressive strain and their transition temperatures could be tuned by strain relaxation. The variation of COO transition with compressive strain can be explained by the strain-dependent lattice deformation by structural phase transition. These results indicate that the lattice degree of freedom plays a crucial role in the stabilization of COO phase in LSMO (Fig. 2.23) [57].

3.3. Strain Effects on the Phase Separation of Manganites

As described in Chapter 1, the manganites tend to be intrinsically inhomogeneous due to the presence of strong tendencies toward phase separation. The phase separation phenomenon in the manganites occurs on various time-scales (from static to dynamic) and length-scales (from glass-like nano to grain-like micron). It is expected that the lattice strain can significantly influence the phase-separated phase of manganites thin films. Theoretical studies show that the phase separation can be ascribed to the intrinsic complexity of a system with strong coupling between the electronic and elastic degrees of freedom, which leads to local energetically favourable configurations and provides a nat-

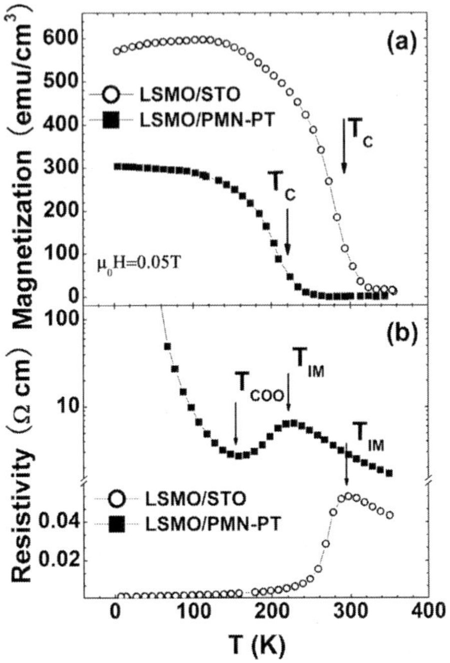

Figure 2.22. (a) The temperature dependent magnetization under a field of 0.05 T for LSMO/STO and LSMO/PMN-PT. (b)The temperature dependent resistivity of LSMO films grown on (001) STO and (001) PMN-PT substrates(Reproduced with permission from Reference [55]. Copyright 2010, AIP Publishing LLC).

ural mechanism for the self-organized inhomogeneities over both nanometre and micrometre scales [58]. Gillaspie *et al.* have studied the strain effect on the $La_{1-x-y}Pr_yCa_xMnO_3$ thin films by epitaxially growing it on $NdGaO_3$, $SrTiO_3$, $SrLaGaO_4$, and $LaAlO_3$ substrates. It is found that the compressive strain from the $SrLaGaO_4$ and $LaAlO_3$ substrates suppresses the long-range charge ordering in these samples and enhances magnetoresistance and magnetic hystereses. Conversely, the tensile strain from the $NdGaO_3$ and $SrTiO_3$ substrates enhances the long-range charge ordering and reduces the magnetoresistance and magnetic hystereses [59]. Dekker *et al.* have applied reversible biaxial strain to $(Pr_{1-y}La_y)_{0.7}Ca_{0.3}MnO_3$ thin film by epitaxially grown on piezoelectric substrates of PMN-PT. A drastic reduction in the resistance was observed upon piezoelectrically controlled release of the tensile strain in the films, which is characterized

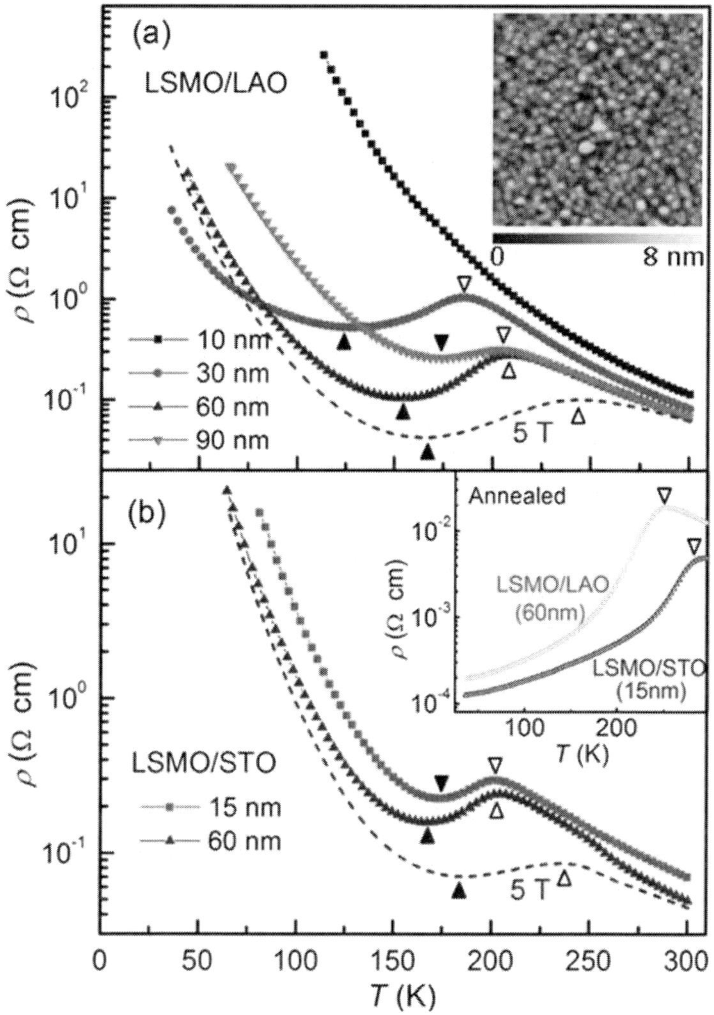

Figure 2.23. Temperature dependence of resistivity for (a) LSMO/LAO and (b) LSMO/STO films. Solid lines with symbol are zero field results and dashed lines are results of the 60 nm LSMO thin films with 5 T field applied. The inset of (a) is a 2×2 μm^2 AFM image of the 60 nm LSMO/LAO film with an RMS of ~2.4 nm. The inset of (b) shows the temperature-dependent resistivity of the annealed LSMO/LAO and LSMO/STO, respectively. The open and closed triangles represent the PMI-FMM transition temperature (T_P) and T_{COO}, respectively(©IOP Publishing. Reproduced with permission from Reference [57]. All rights reserved).

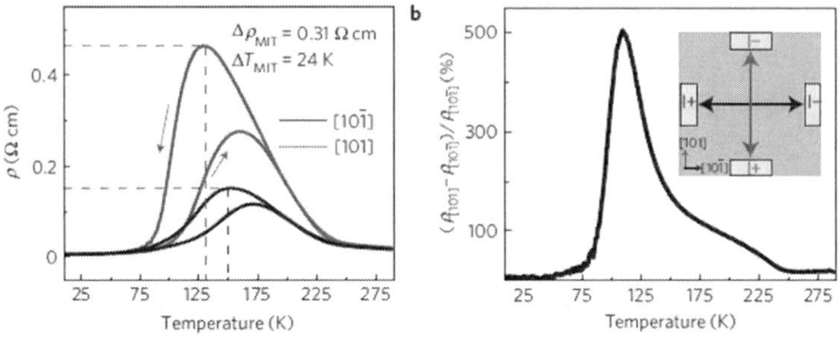

Figure 2.24. a, LPCMO film on an NGO substrate has anisotropic in-plane strain and shows significant differences in both the MIT temperature and the resistivity at the MIT depending on which axis transport is taken across. Arrows indicate cooling and warming curves taken under a 7.5 T magnetic field. b, Relative percentage difference between cooling curves along perpendicular in-plane directions for LPCMO on NGO. (Inset) Diagram showing directions of transport in the LPCMO thin film relative to the in-plane axes(Reprinted by permission from Macmillan Publishers Ltd: Nature Physics (Ref.[51]), copyright (2009)).

by a resistive gauge factor of up to 1800. This indicates that the coexistence of the FM and CO phases is strongly affected by strain with tensile strain suppressing the FM metallic phase [60]. Ward *et al.* have reported transport measurements that show a preferential orientation of electronic phase domains driven by anisotropic long-range elastic coupling between $La_{0.325}Pr_{0.3}Ca_{0.375}MnO_3$ film and $NdGaO_3$ substrate. The anisotropic electronic-domain formation along one axis of a pseudocubic perovskite single-crystal thin-film manganite can be induced by epitaxially locking it to an orthorhombic substrate. Simultaneous temperature-dependent resistivity measurements along the two perpendicular in-plane axes show substantial differences in the metal-insulator transition temperature and extraordinarily high anisotropic resistivity, which indicate that percolative conduction channels open more readily along one axis. These findings suggest that the origin of phase coexistence is much more strongly influenced by strain than by local chemical inhomogeneity (Fig. 2.24) [51].

In addition, phase separation can be induced by strain in the $La_{1-x}(Sr,Ca)_xMnO_3$ that exhibits canonical FMM ground state. Tebano *et al.* have grown different thickness of $La_{0.7}Sr_{0.3}MnO_3$ on $LaAlO_3$ substrates. The mismatch between film and substrate gives rise to an in-plane compressive bi-

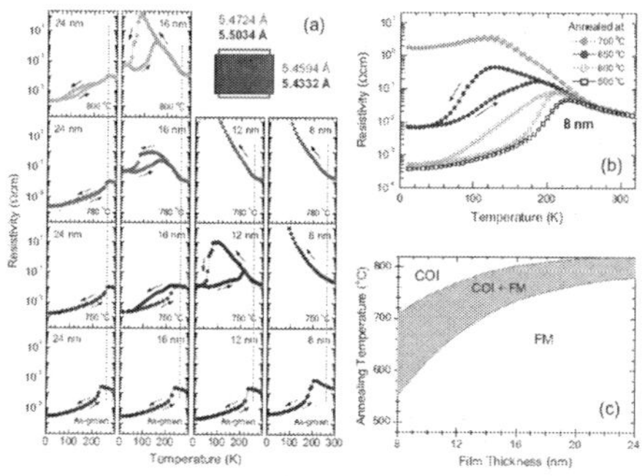

Figure 2.25. (a) $\rho - T$ curves measured from a series of LCMO/NGO(001) films with different film thicknesses and postannealed at different temperatures as denoted. Top right: a schematic of the orthorhombic LCMO films (green) grown on orthorhombic NGO(001) substrates (purple) with their in-plane lattice constants denoted. (b) $\rho - T$ curves from the 8 nm films annealed at 500, 600, 650, and 700°C. (c) Phase diagram of LCMO/NGO(001) films in the plane of film thickness and annealing temperature(Reproduced with permission from Reference [52]. Copyright 2009, AIP Publishing LLC).

axial strain, which partially relaxes in films thicker than 30 unit cells. At the same time, the metal-insulator transition temperature shifts from about 360 K towards lower temperatures with decreasing thickness, eventually reaching an insulating state for the 30 unit cells film. Low-temperature nuclear magnetic resonance spectra show a decrease of the DE contribution with decreasing the thickness, which becomes negligible in the 30 unit cells thick film. The experimental results demonstrate a strain driven competition between two stable phases: the orbital ordered chain-type insulating antiferromagnetic and the orbital disordered metallic ferromagnetic. For intermediate values of the epitaxial strain the local minimum state of the system lies in a gap region between the two stable phases. Such a region has glassy characteristics with coexisting clusters of the two phases. The strain is used as a driving force to span the glassy region [61]. Huang *et al.* have showed that the COOI state can be realized in epitaxial $La_{1-x}Ca_xMnO_3$/$NdGaO_3$ (x=0.3-0.45) thin films via anisotropic strain relaxation. This epitaxial system is special in that there is a negligible average lattice mismatch but a large anisotropic strain in between the film and the substrate. By changing the film thickness, postannealing temperature, along with the doping level for strain relaxation, the COOI phase in the films can be tuned to either melt completely under 1 T, or survive under a high magnetic field of 6 T. The results demonstrate the crucial role of anisotropic strain relaxation in inducing the inhomogeneity in manganites films, thus provide a further understanding of the strain field in manganite physics (Fig. 2.25)[52, 54].

References

[1] A. M. Glazer, *Acta Crystallogr. Sect. B* 28, 3384 (1972).

[2] P. W. Woodward, *Acta Crystallogr. Sect. B* 53, 32 (1997).

[3] P. W. Woodward, *Acta Crystallogr. Sect. B* 53, 44 (1997).

[4] J. M. Rondinelli, and N. A. Spaldin, *Adv. Mater.* 23, 3363 (2011).

[5] P. R. Willmott, and J. R. Huber, *Rev. Mod. Phys.* 72, 315 (2000).

[6] J. T. Cheung, and H. Sankur, *Crit. Rev. Sol. St. Mater. Sci.* 15, 63 (1988).

[7] D. P. Norton, *Mater. Sci. Eng. R* 43, 139 (2004).

[8] Y. Tokura, *Rep. Prog. Phys.* 69, 797 (2006).

[9] A. J. Millis, T. Darling, and A. Migliori, *J. Appl. Phys.* 83, 1588 (1998).

[10] C. Adamo, X. Ke, H. Q. Wang, H. L. Xin, T. Heeg, M. E. Hawley, W. Zander, J. Schubert, P. Schiffer, D. A. Muller, L. Maritato, and D. G. Schlom, *Appl. Phys. Lett.* 95, 112504 (2009).

[11] F. Tsui, M. C. Smoak, T. K. Nath, and C. B. Eom *Appl. Phys. Lett.* 76, 2421 (2000).

[12] M. Ziese, H. C. Semmelhack, K. H. Han, S. P. Sena, and H. J. Blythe, *J. Appl. Phys.* 91, 9930 (2002).

[13] M. Angeloni, G. Balestrino, N. G. Boggio, P. G. Medaglia, P. Orgiani, and A. Tebano, *J. Appl. Phys.* 96, 6387 (2004).

[14] S. W. Jin, G. Y. Gao, Z. Z. Yin, Z. Huang, X. Y. Zhou, and W. B. Wu, *Phys. Rev. B* 75, 212401 (2007).

[15] Y. H. Sun, Y. G. Zhao, M. H. Zhu, B. T. Xie, and W. B. Wu, *J. Appl. Phys.* 112, 023908 (2012).

[16] Y. F. Lu, J. Klein, F. Herbstritt, J. B. Philipp, A. Marx, and R. Gross, *Phys. Rev. B* 73, 184406 (2006).

[17] M. K. Lee, T. K. Nath, C. B. Eom, M. C. Smoak, and F. Tsui, *Appl. Phys. Lett.* 77, 3547 (2000).

[18] D. Dale, A. Fleet, J. D. Brock, and Y. Suzuki, *Appl. Phys. Lett.* 82, 3725 (2003).

[19] C. Thiele, K. Dorr, S. Fahler, L. Schultz, D. C. Meyer, A. A. Levin, and P. Paufler *Appl. Phys. Lett.* 87, 162512 (2005).

[20] C. Thiele, K. Dorr, L. Schultz, E. Beyreuther, and W.-M. Lin *Appl. Phys. Lett.* 87, 262502 (2005).

[21] C. Thiele, K. Dorr, O. Bilani, J. Rodel, and L. Schultz, *Phys. Rev. B* 75, 054408 (2007).

[22] R. K. Zheng, Y. Jiang, Y. Wang, H. L. W. Chan, C. L. Choy, and H. S. Luo, *Phys. Rev. B* 75, 212102 (2007).

[23] R. K. Zheng, Y. Jiang, Y. Wang, H. L. W. Chan, C. L. Choy, and H. S. Luo, *Phys. Rev. B* 79, 174420 (2009).

[24] Z. G. Sheng, J. Gao and Y. P. Sun *Phys. Rev. B* 79, 174437 (2009).

[25] E. J. Guo, J. Gao, and H. B. Lu, *Appl. Phys. Lett.* 98, 081903 (2011).

[26] Y. Konishi, Z. Fang, M. Izumi, T. Manako, M. Kasai, H. Kuwahara, M. Kawasaki, K. Terakura, and Y. Tokura, *J. Phys. Soc. Jpn.* 68, 3790 (1999).

[27] Z. Fang, I. V. Solovyev, and K. Terakura, *Phys. Rev. Lett.* 84, 3169 (2000).

[28] R. P. Borges, W. Guichard, J. G. Lunney, J. M. D. Coey, and F. Ott, *J. Appl. Phys.* 89, 3868 (2001).

[29] Y. H. Sun, Y. G. Zhao, H. F. Tian, C. M. Xiong, B. T. Xie, M. H. Zhu, S. Park, W. D. Wu, J. Q. Li, and Q. Li, *Phys. Rev. B* 78, 024412 (2008).

[30] M. Huijben, L. W. Martin, Y.-H. Chu, M. B. Holcomb, P. Yu, G. Rijnders, D. H. A. Blank, and R. Ramesh, *Phys. Rev. B* 78, 094413 (2008).

[31] A. Tebano, C. Aruta, S. Sanna, P. G. Medaglia, G. Balestrino, A. A. Sidorenko, R. De Renzi, G. Ghiringhelli, L. Braicovich, V. Bisogni, and N. B. Brookes, *Phys. Rev. Lett.* 100, 137401 (2008).

[32] J.-S. Lee, D. A. Arena, P. Yu, C. S. Nelson, R. Fan, C. J. Kinane, S. Langridge, M. D. Rossell, R. Ramesh, and C.-C. Kao, *Phys. Rev. Lett.* 105, 257204 (2010).

[33] Y. Segal, K. F. Garrity, C. A. F. Vaz, J. D. Hoffman, F. J. Walker, S. Ismail-Beigi, and C. H. Ahn, *Phys. Rev. Lett.* 107, 105501 (2011).

[34] Hans Boschker, Jaap Kautz, Evert P. Houwman, Wolter Siemons, Dave H. A. Blank, Mark Huijben, Gertjan Koster, Arturas Vailionis, and Guus Rijnders, *Phys. Rev. Lett.* 109, 157207 (2012).

[35] Y. Tokura, and Y. Tomioka, *J. Magn. Magn. Mater.* 200, 1 (1999).

[36] Y. Tokura, *Rep. Prog. Phys.* 69, 797 (2006).

[37] H. Kawano, R. Kajimoto, M. Kubota, and H. Yoshizawa, *Phys. Rev. B* 53, 14709 (1996).

[38] B. Dabrowski, X. Xiong, Z. Bukowski, R. Dybzinski, P. W. Klamut, J. E. Siewenie, O. Chmaissem, J. Shaffer, C. W. Kimball, J. D. Jorgensen, and S. Short, *Phys. Rev. B* 60, 7006 (1999).

[39] Y. Endoh, K. Hirota, S. Ishihara, S. Okamoto, Y. Murakami, A. Nishizawa, T. Fukuda, H. Kimura, H. Nojiri, K. Kaneko, and S. Maekawa, *Phys. Rev. Lett.* 82, 4382 (1999).

[40] H. Nojiri, K. Kaneko, M. Motokawa, K. Hirota, Y. Endoh, and K. Takahashi, *Phys. Rev. B* 60, 4142 (1999).

[41] S. Uhlenbruck, R. Teipen, R. Klingeler, B. Buchner, O. Friedt, M. Hucker, H. Kierspel, T. Niemoller, L. Pinsard, A. Revcolevschi, and R. Gross, *Phys. Rev. Lett.* 82, 185 (1999).

[42] G. Papavassiliou, M. Pissas, G. Diamantopoulos, M. Belesi, M. Fardis, D. Stamopoulos, A. G. Kontos, M. Hennion, J. Dolinsek, J.-Ph. Ansermet, and C. Dimitropoulos, *Phys. Rev. Lett.* 96, 097201 (2006).

[43] W. Prellier, Ch. Simon, A. M. Haghiri-Gosnet, B. Mercey, and B. Raveau, *Phys. Rev. B* 62, 16337 (2000).

[44] E. R. Buzin, W. Prellier, B.Mercey, C. Simon, and B. Raveau, *J. Phys.: Condens. Matter* 14, 3951 (2002).

[45] Y. Ogimoto, N. Takubo, M. Nakamura, H. Tamaru, M. Izumi, and K. Miyano, *Appl. Phys. Lett.* 86 12513 (2005).

[46] Y. Uozu, Y. Wakabayashi, Y. Ogimoto, N. Takubo, H. Tamaru, N. Nagaosa, and K. Miyano, *Phys. Rev. Lett.* 97, 037202 (2006).

[47] Z. Q. Yang, Y. Q. Zhang, J. Aarts, M.-Y. Wu, and H. W. Zandbergen, *Appl. Phys. Lett.* 88, 072507 (2006).

[48] Y. Z. Chen, J. R. Sun, D. Wei, W. M. Lu, S. Liang, and B. G. Shen, *Appl. Phys. Lett.* 93, 152515 (2008).

[49] Y. Q. Zhang, Z. D. Zhang, and J. Aarts, *Phys. Rev. B* 79, 224422 (2009).

[50] D. Okuyama, M. Nakamura, Y. Wakabayashi, H. Itoh, R. Kumai, H. Yamada, Y. Taguchi, T. Arima, M. Kawasaki and Y. Tokura, *Appl. Phys. Lett.* 95, 152502 (2009).

[51] T. Z.Ward, J. D. Budai, Z. Gai, J. Z. Tischler, Lifeng Yin, and J. Shen, *Nat. Phys.* 5, 885 (2009).

[52] Z. Huang, G. Y Gao, Z. Z Yin, X. X. Feng, Y. Z Chen, X. R Zhao, J. R. Sun and W. B. Wu, *J. Appl. Phys.* 105, 113919 (2009)

[53] F. H. Zhang, Z. Huang, G. Y. Gao, P. F. Chen, L. F. Wang, X. L. Tan, and W. B. Wu, *Appl. Phys. Lett.* 96, 062507 (2010).

[54] Z. Huang, L. F. Wang, X. L. Tan, P. F. Chen, G. Y. Gao and W. B. Wu, *J. Appl. Phys.* 108, 083912 (2010).

[55] J. Wang, F. X. Hu, R. W. Li, J. R. Sun, and B. G. Shen, *Appl. Phys. Lett.* 96, 052501, (2010).

[56] Z. Huang, L. F. Wang, P. F. Chen, G. Y. Gao, X. L. Tan, B. W. Zhi, X. F. Xuan, and W. B. Wu, *Phys. Rev. B* 86, 014410 (2012).

[57] L. Hu, Z. G. Sheng, X. B. Hu, R. R. Zhang, B. Wang, W. H. Song, and Y. P. Sun, *J. Phys. D: Appl. Phys.* 45, 175002 (2012).

[58] K. H. Ahn, T. Lookman, and A. R. Bishop, *Nature* 428, 401 (2004).

[59] D. Gillaspie, J. X. Ma, H.-Y Zhai, T. Z. Ward, Hans M. Christen, E. W. Plummer, and J. Shen, *J. Appl. Phys.* 99, 08S901 (2006).

[60] M. C. Dekker, A. D. Rata, K. Boldyreva, S. Oswald, L. Schultz, and K. Dorr, *Phys. Rev. B* 80, 144402 (2009).

[61] A. Tebano, C. Aruta, P. G. Medaglia, F. Tozzi, G. Balestrino, A. A. Sidorenko, G. Allodi, R. De Renzi, G. Ghiringhelli, C. Dallera, L. Braicovich, and N. B. Brookes, *Phys. Rev. B* 74, 245116 (2006).

In: Manganite-Based Thin Films ...
Editors: Z. Sheng, L. Hu and Y. Sun
ISBN: 978-1-63482-452-1

Chapter 3

ELECTRICAL TRANSPORT IN MANGANITE BASED HETEROJUNCTIONS

*Z. G. Sheng**
High Magnetic Field Laboratory, Chinese Academy of Sciences,
Hefei, People's Republic of China

Abstract

Heterojunction is not only an important structure for electronic and optical devices, but also an ideal system for studying fundamental interface physical phenomena such as band structure, interfacial reconstruction, band-offsets, and interface states. This chapter centers around the electrical transport properties of manganite-based heterojunctions with the perovskite structure. At first, a review of the electric mechanisms in ordinary semiconductor materials is presented and then the manganite-based bilayer and triple layer heterojunctions are discussed in detail.

PACS: 74.20.Mn, 74.25.Dw

Keywords: manganite, heterojunction, electrical transport

*E-mail address: zhigaosheng@hmfl.ac.cn

1. Electrical Transport Mechanisms in Ordinary Semiconductor Materials

Most interesting semiconductor devices usually have two or more different kinds of semiconductors, i.e., heterojucntions.The most important applications of hterojunctions are in photonic devises including semiconductor lasers, photodetectors, and solar cells. There are two kinds of hterojunctions related to semiconductors: junctions consisting tow kinds of semiconductors and junctions consisting of a metal and semiconductor. Moreover, the former one can also be classified into two types from the electrical point of view: p-n heterojunction and isojunctions. The p-n heterojunction consists of two different kinds of semiconductors having different conduction types, whereas the isojunctions is formed by two different semiconductors with same conduction type(such as p-p and n-n heterojunctions). In this section, we briefly review the semicondutor p-n heterojunction.

Let's think about a heterojunction of a p-doped semiconductor with an n-doped semiconductor(Fig. 3.1) in which the two semiconductors are the different. We assume that semiconductor 2 has a wider band gap than 1. The band diagrams of semiconductor 1 and 2 are shown in Fig. 3.2(a).For the separated semiconductor, the position of the bottom of conductor band (E_C) and the top of valence band (E_V) are determined by the chemical bonding of atoms and can be measured or calculated by solving Schroedinger's equation. These energy levels must be measured relatively to some reference level. The reference we have chosen is E_0 (i.e., the vacuum level), the energy of a free electron just outside the neutral materials. The electron affinity, χ, is the energy needed to remove an electron from the semiconductor of an electron located at E_C and make it a free electron.The work function Φ is the distance between E_0 and the Fermi level, E_F.

Once those two semiconductors are contacted with each other, the carriers would be redistributed until equilibrium state is built due to the gap between Fermi levels. As shown in the Fig. 3.1, there are more electrons in $x > 0$ region (n-doped semiconductor side) than in the region $x < 0$ (p-doped semiconductor side). Similarly, there are more holes in $x < 0$ region than in the region $x > 0$. Consequently, electrons will diffuse from the n-side into the p-side and holes will diffuse from the p-side into the n-side. Electrons on the n-side originated from by the ionized donor atoms. When some of the electrons near the interface diffuse into the p-side, they leave behind positively changed donor atoms.

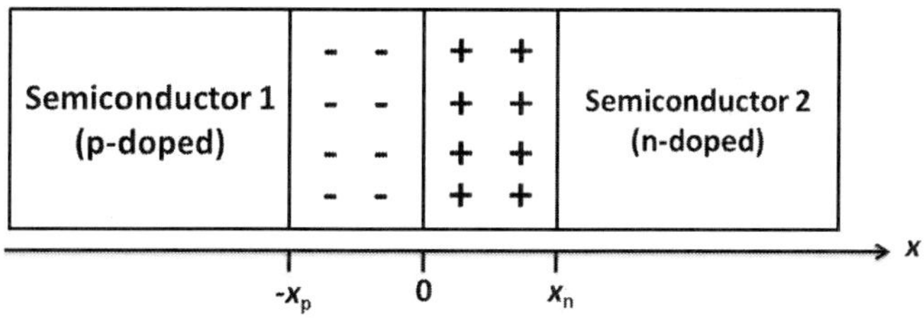

Figure 3.1. Structure of a p-n heterojunction.

Similarly, when the holes near the interface diffuse into the n-side, they leave behind negatively changed acceptor atoms. As a result of these diffusions, a net positive charge density appears in the $x > 0$ region and a net negative charge density appears in the $x > 0$ region as shown in Fig. 3.1. The charged regions ($-x_p < x < x_n$) are called depletion regions because they are depleted of the majority carriers. Moreover, an electric field is generated which points to the negative x-direction (from the donor atoms to the acceptor atoms) due to the charge densities flowing. In equilibrium, as shown in the Fig. 3.2(b), the final structure is characterized by a single Fermi level and a upward band bending in n-side region and downward band bending in p-side region happens. It should be noted that those band bending that occurs inside the depletion regions reflects the presence of an electric field and a corresponding electrostatic potential, i.e., the "built-in" potential V_{bi}. V_{bi} is the sum of band bending in the p-side the n-side, which is equal to the difference in the Fermi levels on the two sides of the junction,

$$-qV_{bi} = E_{F2} - E_{F1} = -q(V_{DN} + V_{DP}) \tag{1}$$

This is because the bands need to bend by this much so that the Fermi level is constant and flat throughout the device in equilibrium.

Considering an abrupt junction with a bias voltage V, the width of the depletion layers in both p- and n-side can be obtained via solving the Poisson's equation in a manner similar to that in the case of homojunctions:

$$x_n = [2\varepsilon_n \varepsilon_p \varepsilon_0 N_A(V_{bi} - V)/qN_D(\varepsilon_n N_D + \varepsilon_P N_A)]^{1/2} \tag{2}$$

and

$$x_p = [2\varepsilon_n \varepsilon_p \varepsilon_0 N_D(V_{bi} - V)/qN_A(\varepsilon_n N_D + \varepsilon_P N_A)]^{1/2} \tag{3}$$

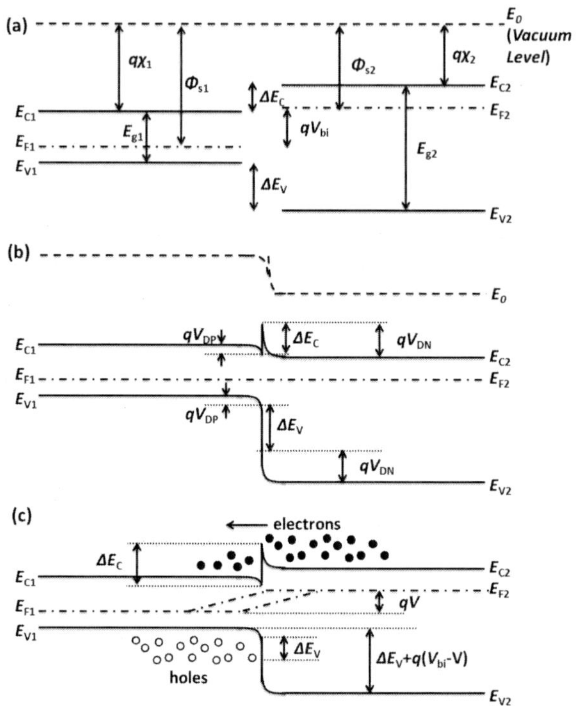

Figure 3.2. (a) The energy band diagram for separated p-doped and n-doped semiconductors. The location for conductor band is measured with respect to a reference energy level E_0, the field free vacuum level. (b)The energy band diagram for contacted a p-n heterojunction.(c)The energy band diagram for a p-n heterojunction under forward bias qV.

where the N_A and N_D are the acceptor and donor concentration in the p- and n-side semiconductors, respectively. And the ε_p and ε_n are the dielectric constants of p- and n-side semiconductors, respectively. Thus, the total width of the depletion layer is

$$W = x_n + x_p = [2\varepsilon_n\varepsilon_p\varepsilon_0(N_D + N_A)^2(V_{bi} - V)/qN_AN_D(\varepsilon_nN_D + \varepsilon_PN_A)]^{1/2} \tag{4}$$

The potential in the p- and n-side semiconductors are related as described by the following equations:

$$N_D\varepsilon_n(V_{DN} - V_n) = N_A\varepsilon_p(V_{DP} - V_p) \tag{5}$$

where V_p and V_n are the biases applied to the p- and n-side semiconductors with V=V_p+V_n. If the dielectric constant of two materials are similar to each other, the applied voltage is mainly biased to the material which has a low doping level.

The depletion layer capacitance per unit area is defined as C=dQ_C/dV, where dQ_C is the incremental increase in charge per unit area upon an incremental change of applied voltage dV. Similar to the homojunctions, the C can be obtained as following equation:

$$C = [q\varepsilon_n\varepsilon_p\varepsilon_0 N_D N_A/(2(\varepsilon_n N_D + \varepsilon_P N_A)(V_{bi} - V))]^{1/2} \tag{6}$$

In general case, people can obtain the value of V_bi through the measurement of $1/C^2 - V$ relationship based on the transformation of above equation as follow:

$$1/C^2 = 2/(q\varepsilon_0)[1/(\varepsilon_n N_D) + 1/(\varepsilon_p N_P)](V_{bi} - V) \tag{7}$$

Let's turn to the current-voltage characteristics of the heteojunctions. Under a bias, the Fermi level of semiconductors would shift as well as the built-in potential. For a forward bais, as shown in Fig. 3.2(c), the V_{bi} can be reduced by the applied voltage and this reduction of V_{bi} will produce the diffusion of electrons from the n-side semiconductor to the p-side semiconductor, while the diffusion of holes is prevented by the large potential barrier, $\Delta E_V + q(V_{bi} - V)$. Similar to the homojunctions, the forward current in a p-n heterojuction is composed mainly of the diffusion current J_{diff}, the surface recombination current J_{SR}, and the generation-recombination current J_{GR}. Consequently, current-voltage characteristics of a p-n heterojuctionthe can be written as

$$I_F = A_{pn}(J_{diff} + J_{GR}) + A_{sur}J_{SR} = I_0 exp[qV/\eta k_B T] \tag{8}$$

where I_0 is the saturation current and η the ideality factor of the junction. For the junction in which the J_{diff} is dominated, the η is close to 1, whereas 1< η <2 if recombination is significant.

Similar analysis can also be applied to the situation with reverse bias. The band diagram in reverse bias is then as shown Fig. 3.3. When V <0 in reverse bias, the electron (hole) density on the p-side (n-side) is pulled below its equilibrium value. This happens because in reverse bias the electric field in the depletion region increases and the drift components of the electron and hole currents exceed the corresponding diffusion components. The large electric field in the depletion region in reverse bias sucks the minority carriers (holes in n-side

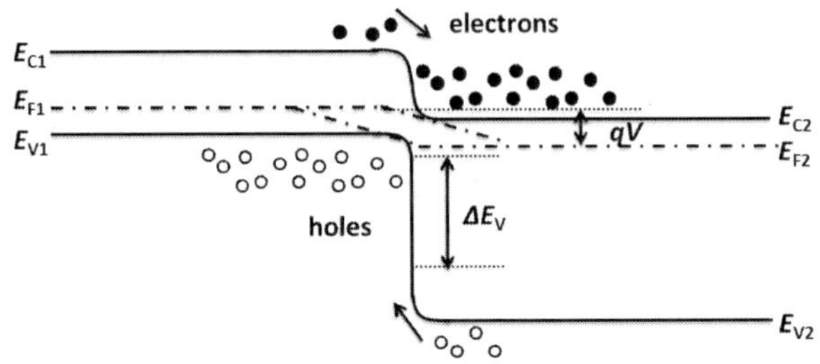

Figure 3.3. The energy band diagram for a p-n heterojunction under reverse bias qV.

and electrons in p-side semiconductor) and transfers them across the depletion region. Consequently, minority carrier generation rate exceeds the recombination rate in the quasi-neutral regions, and this generation of minority carriers on both sides of the depletion region and the subsequent transfer of these generated carriers across the depletion region are responsible for the current in a reverse biased diode. It should be noted that all formulas derived for the forward bias case are also applicable to the reverse bias case, i.e., current-voltage characteristics of the junction under reverse bias can also be described by the Eq.8. It is found that the reverse current approaches $-I_0$ when the reverse bias voltage is large. Figure 3.4 sketches the current-voltage characteristics of the heteojunctions as described above.

2. Manganite Based Bilayer Heterojunctions

The transport mechanism of a pn-heterojunction discussed above is based on the single electron band scenario. This scenario describes the electronic structure of a solid by considering electrons moving independently in an effective periodic potential imposed by ions and other electrons. The periodic potentials make electron dispersion curves discontinues at Brillouin zone boundaries and forms energy bands and gaps in the energy space. Metals contain a partially filled band while insulators' electron band is fully occupied at zero Kelvin. Electrons move freely in the periodic potential and are scattered by impurities in the lattice. This band theory for the electronic states applies rather well and explains most of the electronic properties of metals, insulators, semiconductors or Dirac point metals. In all those cases the independent electrons approach yields a

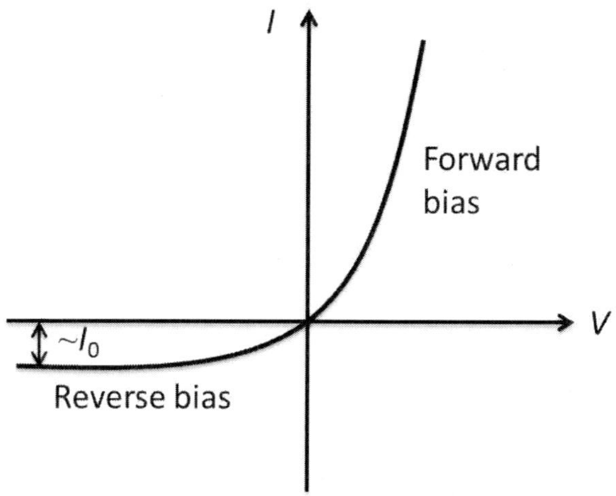

Figure 3.4. The energy band diagram for a p-n heterojunction under reverse bias qV.

weak paramagnetism as all these descriptions do not lift the spin degeneracy of the electronic states. This single electron band theory describes these materials well as the k space construction lifts the site degeneracy of the atomic state by building Bloch states which have different energies and well defined properties under translation. However, this theory neglects the disorder-electron interaction, phonon-electron interaction and electron-electron interaction or treats them as perturbations in the solids. Consequently, in some materials where these interactions become strong, the above approximation would fail to provide a full account of corresponding physical phenomena.

Manganite is one of the typical materials which the single electron band theory is deficient to describe. As mentioned in Chapter 1, the doped manganites with a perovskite structure have strong electronic correlation and show a transition from insulating to metallic state under certain conditions [1]. On the one hand, the correlated electron models should be considered for manganite and manganite-based artificial devices. On the other hand, the rich physical properties caused by the strong correlated electrons, such as metal-insulator transition (MIT), colossal magnetoresistance (CMR), and phase separation (PS), can be utilized in the manganite-based artificial devices which might produce different functionalities from ordinary semiconductor devices. This attracted many researches' attention in last two decades. The bilayer p-n or schottky junctions have been so widely used in semiconductor technology that they are the basic blocks for the semiconductor industry. Though the band theory for semiconductor junctions might be difficult to be applied to manganites due to the strong Coulomb repulsion among electrons, to unify the "Mott transition" in the strongly correlated electron and band gap engineering in a junction would be invaluable, leading to new ways of correlated electronics. There were lots of efforts have been made into this field in the last two decades and both bilayer and trilayer junctions based on manganites have been studied. In this section, the manganite based bilayer heterojunctions and related physical properties will be reviewed.

To build a p-n junction based on manganite, homojunctions composed p- and n-type manganites would come into mind at first. Several groups have made a lot of effort on this issue. Partial substitution of La^{3+} ions by divalent Ca^{2+}, Sr^{2+} and Ba^{2+} in the parent $LaMnO_3$ compound induces both mixed Mn^{3+}-Mn^{4+} valence state and hole-doping. Naturally, electron-doped manganites with a mixed Mn^{2+}-Mn^{3+} state should be obtained in a similar way by using tetravalent cations such as Ce^{4+} for partial substitution of La^{3+} ions [2, 3]. However, up to now both n- [2, 3] and p-type [4] conductivity have been reported for the $La_{1-x}Ce_xMnO_3$ system. Moreover, no clear interface and rectification effect can be found in homojunctions composed with p-type manganite and $La_{1-x}Ce_xMnO_3$ in our experimental results. Thus, on the one hand, it is difficult to build a manganite-based homojunction in current state and additional investigations are necessary to be made in this issue. On the other hand, the manganite based pn-heterojunctions become the key point in this field. Consequently, the hetero-epitaxial growth of manganite films is the first step for the studying of heterojunctions.

Most of doped manganites have a perovskite structure and the lattice constant (pseudo-cubic) is around 3.8 Å. To obtain a manganite film or junction with good crystalline, substrates with similar lattice constant is necessary to be found. There are a lot of oxide single crystals, such as $LaAlO_3$(a=3.821 Å), $NdGaO_3$(a=3.86 Å for pseudo-cubic case), $(LaAlO_3)_{0.3}$-$(Sr_2AlTaO_6)_{0.7}$(LSAT)(a=3.87 Å), $SrTiO_3$(a=3.905 Å), and MgO (a=4.21 Å), which can be utilized for the mangnaite film growth. However, most of these oxides are insulators and not fit for the junctions. Fortunately, the doped $SrTiO_3$ with little Nb or La behaves like a semiconductor and can be used as $n-$type compound in the junction. Remarkable asymmetric $I-V$ characteristics were observed in such a simple junction and the electric field modulation of double exchange ferromagnetism on the basis of a p-n junction was demonstrated in many works [5, 4, 6, 7]. Meanwhile, it was proposed that the bilayer junction is precisely the type of degenerate $p^+ - n$ junction below T_C based on a band diagram raised by Kudo [8], and a p-n one above T_C [5]. It has also been known that the doped $SrTiO_3$ with a little more Nb or La would show a metallic behavior.

Kawai group in Osaka University is one of pioneers in the field of manganite-based bilayer heterojunctions. To form p-n junctions, they deposited the $(La_{0.9}Ba_{0.1})MnO_3$ (LBMO) epitaxial thin films (in thicknesses of 30 and 15 nm)on both Nb-0.01 wt% doped $SrTiO_3(001)$ (STON) and $Sr_{0.99}La_{0.01}TiO_3(001)$ single crystal substrates via the laser molecular-beam epitaxy technique [5, 9]. The high quality LBMO films were deposited with substrate temperature 750 °C and then were annealed at the same temperature in O_2 under 1 atm. In these bilayer junctions, the LBMO was referred as p-type oxide which exhibits ferromagnetic metallic behavior below T_C (~300 K), and the doped $SrTiO_3$ with little Nb or La was used as n-type oxide which behaved like a semiconductor. Remarkable asymmetric $I - V$ characteristics were observed in such a simple junction as shown in Fig. 3.5.

The electron concentration for STON in these junctions was a 6.6×10^{-5}/Ti site (1.1×10^{18} cm^3) at room temperature, confirmed by the Hall coefficient measurement, and hole concentration for LBMO was estimated from the chemical formula as 0.1 per Mn site (an order of 10^{21} cm^3). As proposed by Kudo *et al.*, the doped manganite can be treated as a degenerate semiconductor [8]. They believed that the LBMO/STON junction studied here is precisely the type of degenerate $p^+ - n$ junction below T_C, and a p-n one above T_C, considering the temperature dependence of resistivity and the carrier concentrations. Also, they believed that the conductive electrons should still be correlated by electron repulsion and dynamic Jahn-Teller effects.

Inspired by semiconductor devices and judged from the $J - V$ curves of the LBMO/STON junction, the band picture might be basically applicable to also modulate the carrier concentration in the hole-doped manganites. Figure 3.6 shows the temperature dependence of the LBMO/STON junction resistance, defined as $R_{junction}V_{bias}$=($V_{bias}/J_{interface}$, with various of bias voltage. At a low V_{bias} of 10.6 to 10.8 V, it shows semiconductive property (negative dR/dT). With increasing V_{bias}, $R_{junction}$ is drastically reduced from 3×10^3 to $8\times10^1\Omega$, even within bias voltages from 10.6 to 11.8 V. At V_{bias}=11.0 V, a semiconductor-

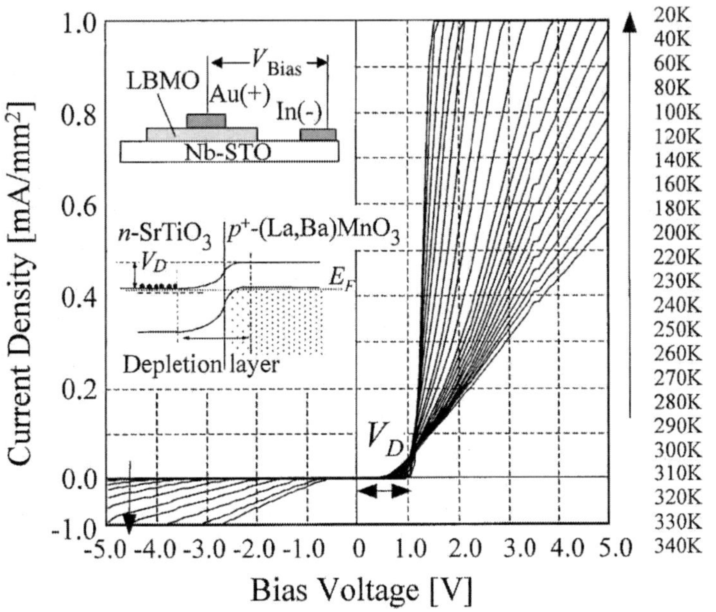

Figure 3.5. Current density-voltage characteristic curves for a $La_{0.9}Ba_{0.1}MnO_3$(LBMO)/Nb-doped $SrTiO_3$ p-n heterojunction perpendicular to the interface at various temperatures (t_{LBMO} = 15 nm). The insets show a schematic illustration of the heterojunction sample and an energy diagram for the p-LBMO n-Sr(Ti, Nb)O_3 heterojunction (Reprinted with permission from Reference [5], Copyright (2002) by the American Physical Society).

to-metal transition appears in $R_{junction} - T$ curves, which is a characteristic feature of strong electron correlation that can be explained by doubt-exchange mechanism. Furthermore, with increasing V_{bias} from 11.0 to 11.8 V, the transition temperature T_P increases gradually from 290 to over 340 K. The change of resistance observed in LBMO/STON juntion includes both the manganite and titanate parts at the interface, but the temperature range of electrical modulation of T_P strongly depends on the intrinsic transport property of the strained LBMO film. Therefore, they concluded that the physical properties of the manganite itself were controlled through carrier modulation by the electric field near the interface.

One point should be noted that the estimated depth of the depletion layer in

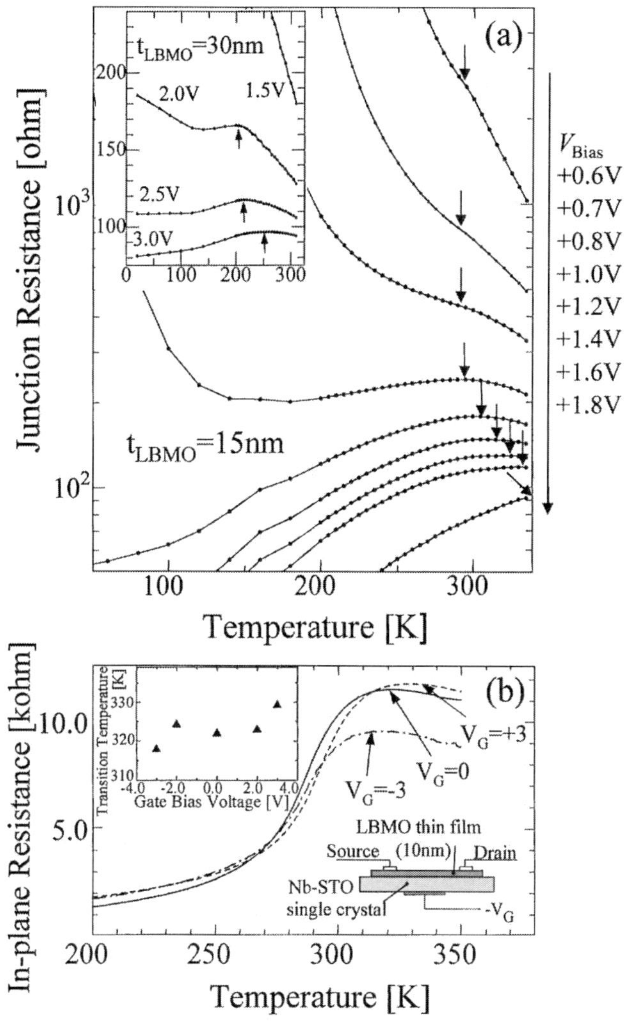

Figure 3.6. (a) Temperature dependence of junction resistance ($R_{junction}$) of a p-(La,Ba)MnO_3 (15 nm)/n-Nb-doped $SrTiO_3$ heterojunction measured under various bias voltages. The arrows indicate metal-insulator transition temperatures. The inset shows that of t_{LBMO} = 30 nm. (b) Temperature dependence of in-plane resistance of (La,Ba)MnO3 layer (t_{LBMO} = 10 nm) in the transistor-type structure under the gate bias voltage (V_G) of 23.0, 0, and 13.0 V, respectively, as shown in the inset illustration. The inset figure shows gate bias voltage dependence of metal-insulator transition temperature (Reprinted with permission from Reference [5], Copyright (2002) by the American Physical Society).

LBMO is less than 1 unit cell (ratio of carrier density $n_{LBMO}/n_{STON} \sim 10^3$) if we use the semiconductor band theory as described in section 1. It is very difficult to explain the electric field effect observed as shown in Fig. 3.6. To explain this, Tanaka *et al.* proposed a scenario about the reduction of effective carrier density as following description. They believed the metal-insulator transition in the transition metal oxides is anomalous. Near the insulator region, most carriers are localized by strong Coulomb repulsion, so that the number of conductive carriers is greatly reduced even if formal carrier concentration is of the order of 10^{21} cm^3. Furthermore, electron-lattice (Jahn-Teller) interaction enhances this localization tendency. This scenario can be verified to some degree by the insulative $La_{1-x}Sr_xMnO_3$ ($0< x <0.175$) even with a formal carrier of the order of 10^{21} cm^3. Therefore the depletion layer of LBMO appears to become much deeper than that of an ordinal semiconductor, due to electron localization by strong electron correlation (Coulomb repulsion, Jahn-Teller effect). With this in mind, it is easy to understand the electric field effect in LBMO/STON junction. According to the phase diagram of hole-doped manganites [1], enhancement of the hole concentration enhances ferromagnetic double exchange interaction, leading to the decrease of resistivity and the increase of T_P. In the LBMO/STON junction, the effective hole concentration near the interface is enhanced with increasing of V_{bias}, leading to the change of electromagnetic properties in the LBMO layer with a enhancement of T_P.

One of the most fascinating characteristics of the manganites is their ability to control both magnetic and electrical properties drastically by the magnetic fields. By applying a magnetic field of 5 T, $R_{junction}$ of LBMO/STON was reduced in all temperature range from 300 to 10 K (t_{LBMO} = 30 nm), as shown in Fig. 3.7. With decreasing temperature, this magnetoresistance effect increased monotonically. This MR effect was enhanced from 5% to 30% at 10 K, while decreasing V_{bias} from 3.0 to 1.0 V. The magnetoresistance of semiconductive STON is almost negligible comparing with that of magnetoresistive LBMO, so the change of junction resistance is mainly dominated by LBMO. As presented in this work, both electric field and magnetic field can modulate the transport properties of manganite-based bilayer junction. It may open up a window for the correlated electron devices which can switch rich functionality of complex oxides, such as ferromagnetism, superconductivity, and more.

Bilayer manganite-titanate junction is the most simple building blocks for artificial structures. As mentioned above, undoped $LaMnO_3$ is a correlated-electron insulator, while undoped $SrTiO_3$ is a band insulator. Although it is

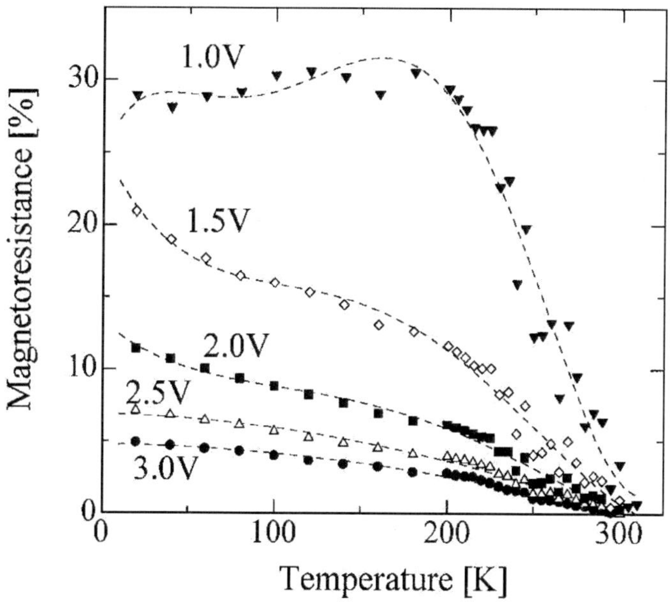

Figure 3.7. Temperature dependence of magnetoresistance at the magnetic field of 5 T measured under various bias voltages of the LBMO/STON heterojunction t_{LBMO} = 30 nm. The MR ratio is defined as $[R_{junction}(H{=}0$ T)-$R_{junction}(H{=}5$ T)/$R_{junction}(H{=}0$ T)$\times 100\%$ (Reprinted with permission from Reference [5], Copyright (2002) by the American Physical Society).

an open question about how to model the interface between a doped correlated insulator and doped band insulator, a lot of efforts have been made on this issue. Significant junction magnetoresistance have been observed in these junctions [10, 7, 11, 12, 13].To reveal the transport mechanism of those bilayer manganite-titanate junctions in detail, several kinds of analysis have been done. Among them, junction current-voltage ($I - V$) characteristics measurement is a powerful technique to study interface electronic structure since (1) it can be performed even under external magnetic field and (2) it sensitively probes the thin interface. In 2007, Susaki *et al.* analyzed the temperature dependence of $I - V$ characteristics of single interface $La_{0.7}Sr_{0.3}MnO_3$(LSMO)/STON and $La_{0.7}Sr_{0.3}MnO_{3-\delta}$(LSMO-$\delta$)/STON junctions and the transport mechanisms have been discussed [13]. Since the gap size of undoped $LaMnO_3$ is much smaller than that of undoped $SrTiO_3$, they treated the LSMO/STON junction approximately as a metal-(n-type) semiconductor Schottky junction, where only electrons contribute to the junction transport process. The thermionic emission current density [current density $J \equiv I/$(junction area) across the Schottky barrier is given by [14]

$$J \cong J_S(T)exp[qV/nk_BT] \tag{9}$$

for $V \gg k_BT/q$. Here, $J_S(T)$ is the saturation current density, q is the electronic charge, n is the ideality factor (equal to unity in a purely thermionic emission case), and k_B is the Boltzmann constant. Panels (a) and (b) of Fig. 1 show that the current is linear in the forward-bias voltage on a semilogarithmic scale and the slope of $I - V$ characteristics increases in the temperature range from 400 to 100 K, which is qualitatively consistent with thermionic emission current, while it is almost independent of the temperature below 100 K in both LSMO and LSMO-δ junctions.

By fitting Eq.9 to the observed $I - V$ characteristics shown in Fig.3.8, the saturation current density $J_S(T)$ can be obtained. Then, the temperature dependence of the barrier height ϕ_{bi} can be extracted by following equation.

$$J_S(T) = A^*T^2exp[-(q\phi_{bi})/k_BT] \tag{10}$$

where they set the effective Richardson constant A^* as 156 Acm^{-2} K^{-2}, which corresponds to the effective mass m^*/m_0=1.3 for Nb:$SrTiO_3$ [15]. The temperature dependence of $J_S(T)$ and ϕ_{bi} for both junctions at H=0 were shown in Fig. 3.9. In both junctions, the barrier height ϕ_{bi} gradually decreases as the temperature is lowered and is finally reduced to an unphysically small value of $\sim$0.05 V at 10 K.

The temperature dependence of $I - V$ characteristics shown in Fig. 3.8 and 3.9 reveals that the transport properties deviated from ideal thermionic emission model especially at low temperatures. They attribute this deviation to the (thermally assisted) tunneling process. At low temperature, electrons between the Fermi level (E_F) and conduction-band bottom tunnel to the metal under forward bias (direct tunneling), while at higher temperature, they are firstly thermally excited to an energy level between E_F and the top of the barrier and then they tunnel to the metal (thermally assisted tunneling). To facilitate understanding those transport process well, Figure 3.10 schematically illustrate both the direct and thermally assisted tunneling processes in comparison with the thermionic emission process. Within the semiconductor band model, the forward-bias $I - V$ characteristics in the direct tunneling regime are expressed

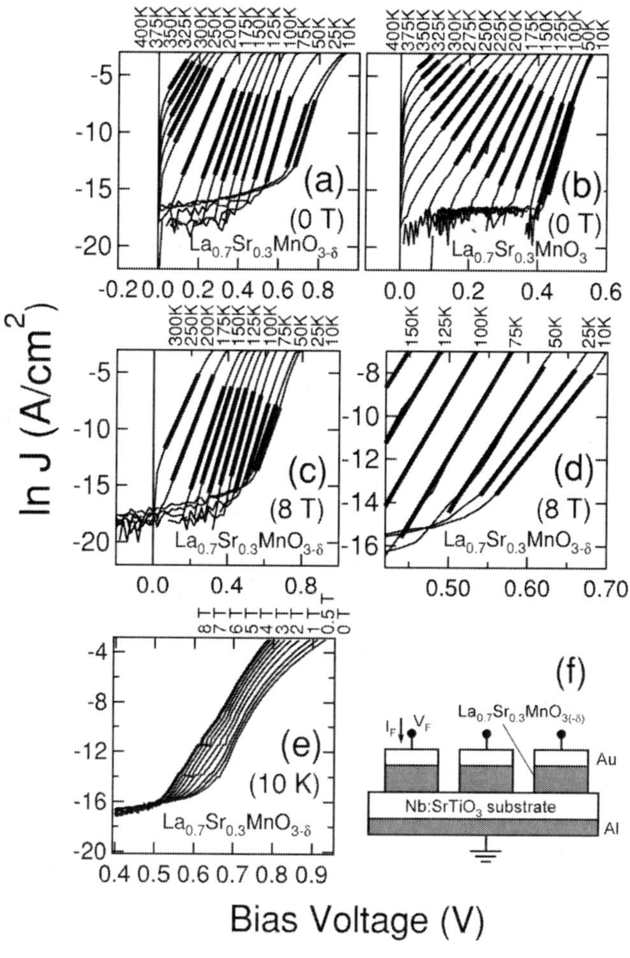

Figure 3.8. Forward-bias $I - V$ characteristics for the (a) oxygen deficient and (b) oxygen-stoichiometric junctions at 0 T. Characteristics for the oxygen-deficient junction at 8 T are shown in (c) as well as in (d) on an enlarged scale. Bold lines are linear fitting on a semilogarithmic scale. (e) Magnetic field dependence of the forward-bias $I - V$ characteristics for the oxygen deficient junction at 10 K. (f) Schematic layout of the junction device for $I - V$ and capacitance measurements (Reprinted with permission from Reference [13], Copyright (2007) by the American Physical Society).

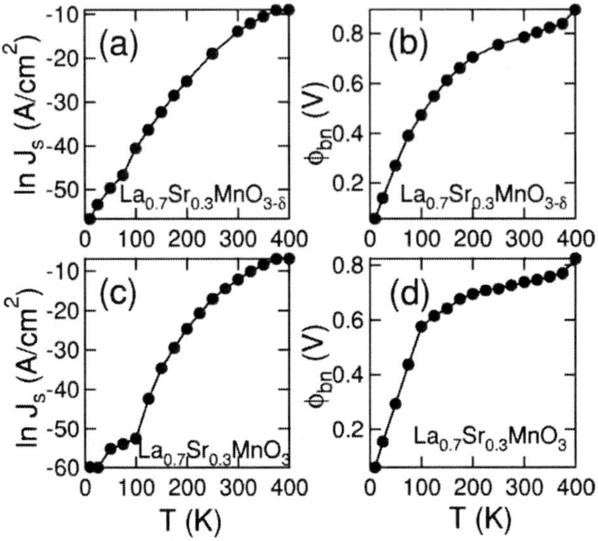

Figure 3.9. (a) Saturation current density $J_S(T)$ and (b) barrier height ϕ_{bi} as a function of temperature for the oxygen-deficient junction at 0 T. $J_S(T)$ and ϕ_{bi} for the stoichiometric junction at 0 T are plotted in panels (c) and (d), respectively (Reprinted with permission from Reference [13], Copyright (2007) by the American Physical Society).

as

$$J = J_S(T)exp[qV/E_{00}] \tag{11}$$

where $J_S(T)$ depends weakly on temperature as $cT/sin(cT)$, in which c is a constant. E_{00} is a temperature-independent parameter. As Eq.11 shows, E_{00}/q corresponds to the inverse of slope in ln $J-V$ plots, and E_{00}=15.6 meV for the oxygen-deficient junction and 9.9 meV for the stoichiometric junction according to the slope observed at 10 K for H=0. On the other hand, in the thermally assisted tunneling regime, the energy of tunneling electrons is distributed around E_m as shown in Fig.3.10, the peak position of the energy distribution of tunneling electrons, and both E_m and the distribution energy width increase as the temperature increases.

Based on the analysis above, they concluded that the temperature-dependent $I-V$ characteristics of the stoichiometric LSMO/STON junctions were fully consistent with the (thermally assisted) tunneling model. (Thermally assisted)

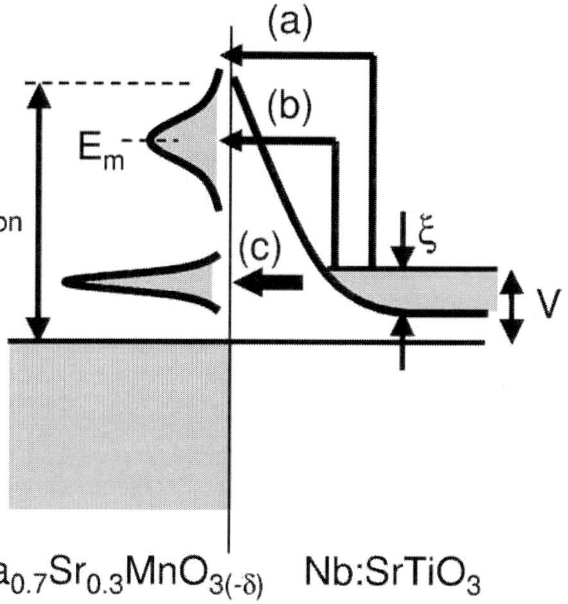

Figure 3.10. Schematic potential profile for $La_{0.7}Sr_{0.3}MnO_{3-(\delta)}$/Nb:$SrTiO_3$ Schottky junction under the forward bias of V. (a) Thermionic emission, (b) thermally assisted tunneling, and (c) direct tunneling processes from Nb:$SrTiO_3$ to $La_{0.7}Sr_{0.3}MnO_{3-(\delta)}$ are also schematically shown (Reprinted with permission from Reference [13], Copyright (2007) by the American Physical Society).

Tunneling is relevant when the Schottky barrier is relatively thin. However, it is found that the $I - V$ characteristics of the magnetoresistive oxygen-deficient $La_{0.7}Sr_{0.3}MnO_{3-\delta}$ junction are quite unusual for $H \neq 0$, strongly deviating from conventional thermally assisted tunneling as shown in Fig. 3.8(c). They suggested that the deviation arises from the changes of the Schottky barrier caused by magnetic field, originating from double-exchange coupling to the electronic structure of the interface.

Almost at the same time, our group also did similar work about the Ce-doped $LaMnO_3$-titanate junction. As predicted and proved by some group [2, 3], the tetravalent Ce^{4+} cations doping can produce a n-type manganite. Then, the n-n junction could be realized in (LaCe)MnO_3/STON interface. We made the $La_{0.7}Ce_{0.3}MnO_3$(LCEM)/STON heterojunction by growing a LCEM film on 3×5 mm^2 STON crystal with the (001) orientation using laser ablation. The

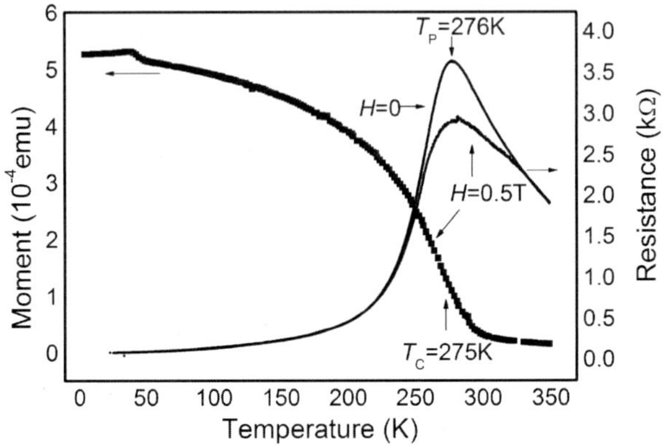

Figure 3.11. In-plane resistance and magnetization of $La_{0.7}Ce_{0.3}MnO_3$ film as functions of temperature in the $La_{0.7}Ce_{0.3}MnO_3$/STON heterojucntion (Reprinted from Reference [20], Copyright (2006), with permission from Elsevier).

film was $\sim$ 500 Å in thickness. To improve crystal quality and keep the oxygen stoichiometry, the sample was maintained at 750 °C for 20 min in flowing O_2 gas after deposition. The temperature dependence of the resistance and magnetization of LCEM film are plotted in Fig. 3.11. It is found that the Curie temperature T_C of LCEM film is higher ($\sim$275 K) than that of its corresponding bulk material ($\sim$236 K) [17]. Such a phenomenon, we observed, was fully consistent with a previous report [18], in which it was found that a compression of unit cell in the $a - b$ plane due to epitaxial strain. It was suggested that the reduction of the unit-cell volume due to strain effect would enhance the transfer integral of electron hopping and thus enhance T_C. Additionally, as proved by many experiments [32, 5], the depletion layer in the manganite-based heterojunction may be intrinsically inhomogeneous because of strains. That is to say, there exist some insulating phases except for metallic phase in the depletion layer caused by the interface strain even below T_C.

Fig. 3.12 shows the typical results of $I - V$ measurements measured at 320 K. It indicates that the $I - V$ curve shows good rectifying behavior. The voltage exhibits a tendency to saturate at V_d for positive bias current and V_b for the opposite direction bias. It has been proved that the band diagram description is applicable to the manganites [8]. In this scenario, V_d is diffusion voltage be-

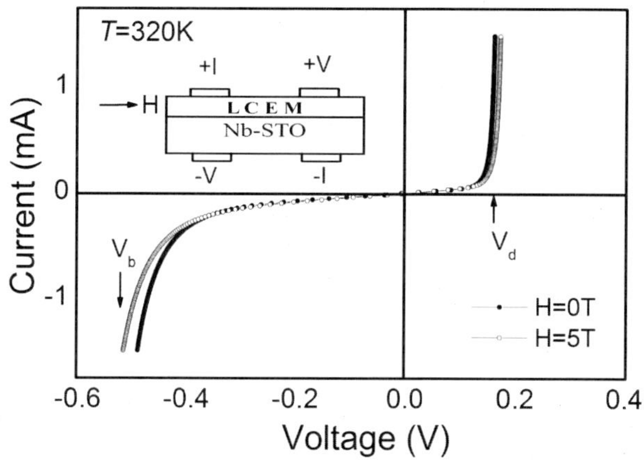

Figure 3.12. $I - V$ characteristics of $La_{0.7}Ce_{0.3}MnO_3$/STON heterojucntion at 320 K with and without magnetic field 5 T which is parallel to the $La_{0.7}Ce_{0.3}MnO_3$ film plane. The inset shows a schematic illustration of electrode setting (Reprinted from Reference [20], Copyright (2006), with permission from Elsevier).

tween the LCEM and STON, which appears when a semiconductor is brought into contact with the other with a different carrier type of band structure. And V_b is a breakdown voltage that when being exceeded by the applied electric field, sudden current increase usually occurs. In addition to the two current rushes at V_d and V_b, respectively, which are typical features of bipolar heterojunctions, the $I - V$ curve under zero and 5 T magnetic fields show a significant magnetoresistance (MR) within the whole measured temperatures, and the typical results at 320 K are presented in Fig. 3.12.

Fig. 3.13 shows V_d and the corresponding differentiate as function of temperature. With increasing temperature, the V_d decreases non-linearly. Especially, a sharp change occurs in the vicinity of Curie temperature of LCEM/STON film. This phenomenon reveals the existance of a close correlation between the transport properties of LCEM film and LCEM/STON junction. To investigate variation of energy gap of LCEM/STON heterojunction further, Vd under an external magnetic field of 5 T that is parallel to LCEM film plane has been studied. The application of magnetic field causes an obvious shifting of Vd T curve to the higher temperature as shown is Fig. 3.13(b). It shows that there exists a minimum value at the temperature of 246 K, and 264 K in differential curves of Vd under the applied fields of 0 T, and 5 T, respectively. The maximum change of Vd occurs at T=275 K, where Vd increass from 0.24 V for H=0 to 0.30 V for H=5 T, and the relative variation Vd (defined as [Vd(5T)-Vd(0T)]/Vd(0T)) is as much as 28%. Considering that the behavior of STON is magnetic field independent, we attribute the large shifting of $V_d - T$ curve under applied fields to the variation in energy band of LCEM electrode resulting from the change of magnetic order induced by the external magnetic field.

It is well known that the junction "breaks down" and conducts a very large current when a sufficiently high electric field is applied to a bipolar heterojunction. Therefore, the breakdown voltage (V_b) is an important parameter of heterojunctions, and its variation enables us to investigate the transport mechanism of the LCEM/STON heterojunction. Fig. 3.14 presents V_b as a function of temperature under an external magnetic field oriented parallelly to LCEM film plane. It is obvious that V_b(T) curve in zero-magnetic field is quite different from V_d(T) curve as shown in Fig.3.13(a) and V_b(T) curve has a local maximum appearing near 175 K, i.e., V_b increases with increasing temperature below 175 K, and decreases with increasing temperature above 175 K. Moreover, the peak temperature has a tendency shifting to higher temperatures as the magnetic field is exerted on the heterojunction.

The different temperature dependence of V_b below and above 175 K may originate from the different reverse transport mechanism. In the framework of the traditional semiconductor theory, when the electric field at the junction approaches a threshold value, significant current begins to flow (break down) by means of band-to-band tunneling process. The tunneling current density in reverse bias condition is given by

$$J_R \propto \varepsilon V E_g^{1/2} exp[-\alpha E_g^{3/2}] \tag{12}$$

where ε is the electric field of the junction, V is the applied voltage, and α is the constant for a given heterojunction. In the LCEM/STON heterojunction, the resistance of LCEM film and STON substrate can be negligible compared with the resistance of depletion layer of the junction. Therefore, it is reasonable to assume ε is equal to applied voltage V. For a given break down current J_{Rb}, equation 12 can be written as

$$\varepsilon V_b \approx V_b^2 \propto J_{Rb} E_g^{1/2} exp[\alpha E_g^{3/2}] \tag{13}$$

Since the E_g of LCEM/STON decreases with increasing temperatures as proved by the variation of V_d, the V_b in the heterojunction because the tunneling effect has a negative temperature coefficient, that is to say, the V_b decreases with increasing temperatures. This is because a given breakdown current J_{Rb} can be reached at small reverse voltage (or fields) and higher temperatures. This temperature dependence is quite different from ionization mechanism, which has a positive temperature coefficient. As to our experiments, the breakdown current caused by tunneling effect may be dominant when the temperature is above 175 K. Similar phenomenon was also observed in other manganite-STON heterojunctions [19], which indicates reverse tunneling transport mechanism could be a common feature in manganite-STON heterojunctions. On the other hand, the V_b has a positive temperature coefficient below temperature of 175 K that is contrary to the one above temperature of 175 K. It indicates that tunneling effect is not a dominant reverse transport mechanism any more in LCEM/STON heterojunction below 175 K. It is worthy of noting that the critical temperature (T=175 K) at which the crossover of transport mechanism takes place is far below the Curie temperature (T_C=275 K) of the LCEM/STON heterojunction.

The temperature dependent transport mechanism of LCEM/STON heterojunction is very important for underlying practical application. To make a primary explanation, we prefer to attribute this temperature dependent property to the co-work of E_g and insulator parts in the depletion layer which may exist at the interface as a dead layer, such as the CO state or Jahn-Teller polarons. As mentioned before, there exist epitaxial strain between LCEM film and STON substrate and it may result in the apearence of the phase separation in the depletion layer of the heterojunction. And the phase separation in mangnites even exists far below T_C as suggested by recent theoretical and experimental studies [1]. That is to say, those insulator parts in the depletion layer may affect electric transport of the heterojunction even below T_C. With temperature decreasing, the energy band gap E_g increases and provides a favorable condition for carriers to tunnel. While the insulator parts in depletion layer would play a contrary role in the tunneling effect. The fractions of both FM metallic and FM insulating phases can be changed by changing temperatures and applied magnetic fields, which results in the variation of the depletion layer and the transport mechanism. Based on our experimental results, it seems to be reasonable that there exists FM insulator layer in the depletion layer that composes a barrier to prevent carriers from moving except tunneling even far below T_C. With decreasing temperatures, FM insulating layer becomes thinner, whereas FM metallic parts becomes thicker. When temperature decreases to a critical value, FM metallic channels form and it is against to tunneling effect. For the present LCEM/STON heterojunction, the temperature of percolation threshold locates in the vicinity of 175 K. In this scenario, external magnetic fields will contribute to enlarge FM metallic parts and enhance the transition temperature. Fortunately, this relation has been proved by our experiment as seen it Fig. 3.14.

As mentioned previously, though the n-type manganite was expected with tetravalent Ce^{4+} cations doped, the sign of majority carrier in (LaCe)MnO_3 is still in controversy. In our transport results shown above in (LaCe)MnO_3-based junction, it is still difficult to solidify this issue. However, undoubtedly, both the divalent and tetravalent ion-doping in $LaMnO_3$/STON heterojunctions provide us platforms to recognize the correlated electron interfaces in detail.

At the beginning of the studying of manganite-titanate heterojunctions, the $I-V$ characteristics show rectification but the breakdown voltage with reverse bias condition is lower [4, 20]. With development of junction fabrication techniques, the electrical transport properties of manganite-titanate heterojunctions become more stable and reproducible. With developing of junction techniques, Sawa *et al.* fabricated epitaxial $Pr_{0.7}Ca_{0.3}MnO_3$ /$SrTi_{0.9998}Nb_{0.0002}O_3$ (PCMO/STON) junctions, in which the epitaxial PCMO films (100 nm thick) were deposited on (100) STON substrates by a pulsed laser deposition at a substrate temperature of 700 °C under an oxygen pressure of 100 mTorr. As a reference, the $La_{0.7}Sr_{0.3}MnO_3$/$SrTi_{0.9998}Nb_{0.0002}O_3$ (LSMO/STON) junctions have been made with same conditions. Ohmic contact to the semiconducting PCMO film was made by a 20 nm thick $SrRuO_3$(SRO) layer subsequently deposited on the PCMO layer under the same conditions. To provide ohmic contacts to the metallic SRO and LSMO films, Au films were $ex situ$ deposited. Contacts to the STON substrates were provided by the Ag paste on the backside. The Au/LSMO and Au/SRO/PCMO structures were patterned into mesas of 200×200 mm^2 by conventional photolithography and Ar-ion milling. The $J-V$ characteristics were measured by two-point contacts method. In their paper, the positive bias is defined as the current flows from Nb:STO substrate to Au electrode.

Figures 3.15(a) and 3.15(b) show the current density $J-V$ characteristics of PCMO/STON and LSMO/STON junctions, respectively, having clear rectification. As mentioned in section 1, both p-n and Schottky diodes follow the same relation of $\ln J_F \sim qV/nk_BT$. Actually, $\log_{10}J_F$ of both junctions changes linearly with V at low bias region for T >100 K, as shown in the insets of Figs. 3.15(a) and 3.15(b). From the slopes of the $J_F - V$ curves, the ideality factor n can be deduced as n=1.05-1.10 for both junctions above T=150 K. However,

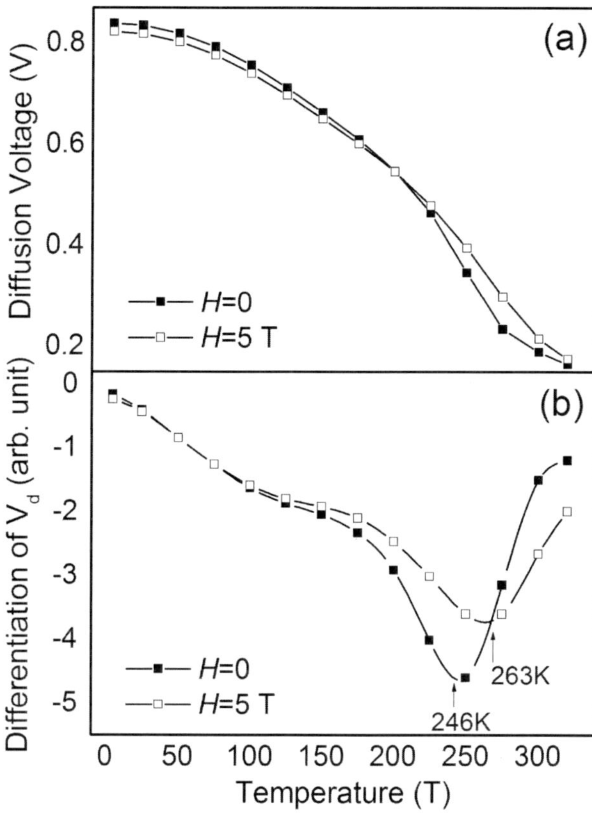

Figure 3.13. (a) Diffusion voltage (V_d) of $La_{0.7}Ce_{0.3}MnO_3$/STON heterojucn-tion as function of temperature measured under H=0 and 5 T. (b) The differentiation of V_d as a function of the temperature for H=0 and 5 T (Reprinted from Reference [20], Copyright (2006), with permission from Elsevier).

below 100 K, $\log_{10} J_F$ relation deviates from the systematic trend at higher temperatures for both junctions. This is presumably due to the onset of quantum paraelectricity in the depletion layer of STON [21], where the dielectric constant strongly depends on the bias voltage and the bias voltage dependence of the depletion layer width deviates from the ones expected for p-n and Schottky diode models. In addition, the phase transition of PCMO from paramagnetic insulator to charge-ordered insulator around 200 K may cause further deviation

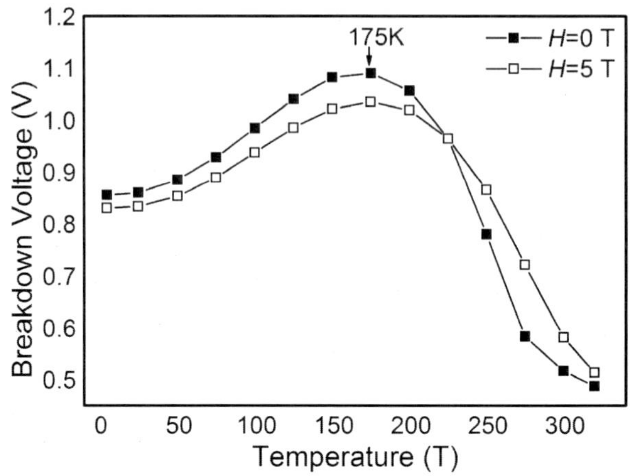

Figure 3.14. Breakdown voltage (V_b) of $La_{0.7}Ce_{0.3}MnO_3$/STON heterojucntion as function of temperature measured under H=0 and 5 T (Reprinted from Reference [20], Copyright (2006), with permission from Elsevier).

[22, 23]. For the PCMO/STON diode, the suppression of J_F below 100 K or the shift of turn-on voltage may be due to the change of electronic states of PCMO upon this phase transition.

In the reverse bias (positive biased in Fig. 3.15), the leakage current density (J_L) of both diodes is less than 1×10^6 A/cm^2 at V=5 V for the measured temperature range (T=10-300 K). At T=300 K, J_L at V=100 V is about 2×10^2 A/cm^2 and 7×10^2 A/cm^2 for the PCMO/STON and LSMO/STON diodes, respectively. For the PCMO/STON diodes, an apparent breakdown is not observed for the measured temperature range. However, for the LSMO/STON junction, J_L increases and an apparent breakdown voltage V_D decreases as T decreases. That V_D decreases with the decreasing of T was already observed in selenium rectifiers and explained by the suppression of Schottky barrier height ϕ_B due to the image force effect and the temperature dependence of the electron mobility (m) [24, 25].

It is noted that the $J-V$ characteristics of both diodes exhibit hysteresis in the reverse bias region but not in the forward bias region. Such hysteretic $J-V$ characteristics have been observed in SRO/STON heteroepitaxial Schottky junctions [26]. In the SRO/STON case, hysteresis are observed in both forward and reverse bias regions, and this hysteretic property leads to a reversible resistance

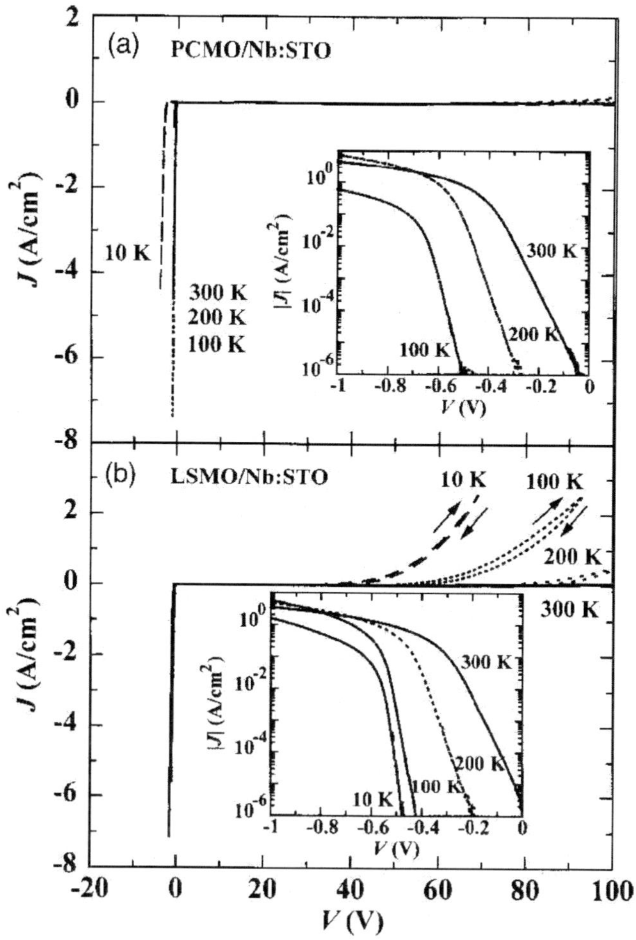

Figure 3.15. $J - V$ characteristics of (a) PCMO/STON and (b) LSMO/STON diodes. The insets show $J - V$ characteristics drawn in semilogarismic current scale in the forward (negative) bias region (Reproduced with permission from Reference [11]. Copyright 2005, AIP Publishing LLC).

switching induced by pulsed voltage stress. However, the PCMO/STON and LSMO/STON diodes do not exhibit such a resistance switching effect.

In a p-n anisotype heterojunction model, current density J can be expressed as

$$J = J_{pn}(1 - V/V_{BI})[exp(qV/k_BT) - 1] \quad (14)$$

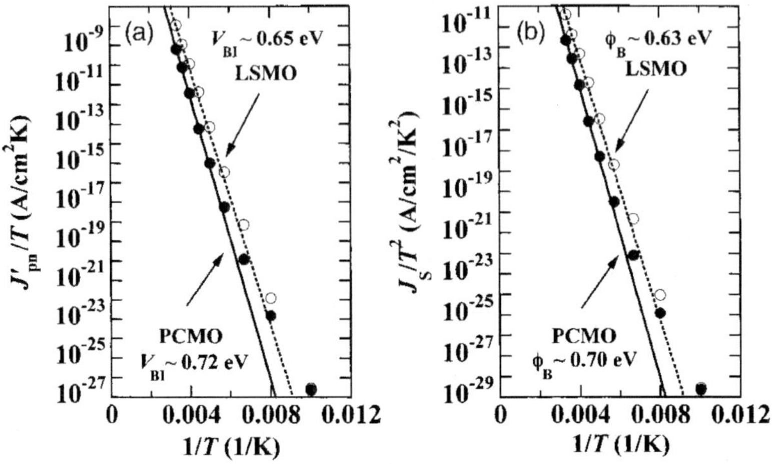

Figure 3.16. (a) $J'_{pn}/T - 1/T$ and (b) $J'_{pn}/T^2 - 1/T$ characteristics for the PCMO/STON and LSMO/STON diodes. The lines are theoretical fits using (a) $\ln J'_{pn}/T \propto qV_{BI}/k_BT$ and (b) ln $J_S/T^2 \propto q\phi_B/k_BT$ (Reproduced with permission from Reference [11]. Copyright 2005, AIP Publishing LLC).

$$J_{pn} = A^*qTV_{BI}exp(-qV_{bi}/k_BT)/k_B \tag{15}$$

where A^* is the Richardson constant, V_{BI} is the built-in potential of p-n junction, and k_B is the Boltzmann constant. The Richardson constant can be estimated from $A^* = 4\pi m^* k_B/h^3$, where h is the Planck's constant and m^* the effective mass of electron. When m^* is taken as $1.3m_0$ with m_0 being free electron mass, A^* is evaluated as 156 A cm^{-2}K^{-2} for STON [21]. The forward bias current density (J_F) on qV/k_BT is then expressed approximately as

$$J_F = J'_{pn}exp(qV/nk_BT) \tag{16}$$

because the last exponential term in Eq. 14 becomes dominant at $V > k_BT/q$. Here, n is the ideality factor. Assuming that both diodes are p-n diodes, we evaluate built-in potential of the junction V_BI from the temperature dependence of J_{pn}. From Eqs.15 and 16, the temperature dependence of J'_{pn} extracted from $J_F(V)$ should satisfy the condition that ln $J'_{pn}/T \propto qV_{BI}/k_BT$, when n is close to unity. As shown in Fig. 3.16(a), $J'_{pn}/T - 1/T$ curves of both diodes follow straight lines as predicted above T=150 K. Here, V_{BI} evaluated from the $J - V$ characteristics in the forward bias region is defined as V^J_{BI},

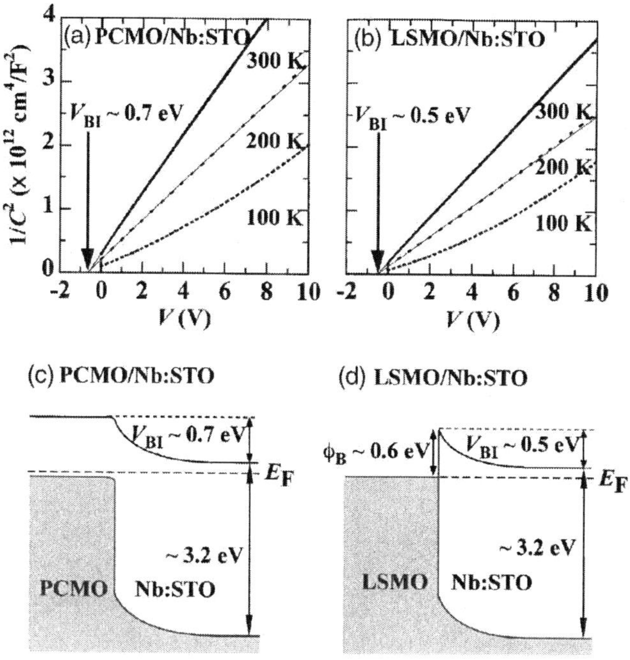

Figure 3.17. $1/C^2 - V$ characteristics measured at a frequency of 10 kHz for (a) PCMO/STON and (b) LSMO/STON diodes. Possible band diagrams of (c) PCMO/STON p-n diode and (d) LSMO/STON Schottky diode (Reproduced with permission from Reference [11]. Copyright 2005, AIP Publishing LLC).

and the V_{BI}^J values for the PCMO/STON and LSMO/STON are 0.72 and 0.65 eV, respectively. Another way to deduce V_{BI} is by means of $C - V$ characteristics in the reverse bias region. For both p-n and Schottky diodes, $1/C^2$ linearly varies with V, and the V at which $1/C^2$ intercepts the V axis gives V_{BI}; $1/C^2 \sim (V_{BI} - V)$. Figures 3.17(a) and 3.17(b) show $1/C^2 - V$ characteristics for the PCMO/STON and LSMO/STON diodes. Above T=200 K, $1/C^2$ changes linearly with V. However, at T=100 K, $1/C^2$ shows nonlinearity with V because the dielectric constant in the depletion layer of the STON changes with V. Here, V_{BI} evaluated from the $C - V$ characteristics is defined as V_{BI}^C. The V_{BI}^C values for the PCMO/STON and LSMO/STON diodes are almost constant above T=200 K, and are ~0.7 and ~0.5 eV, respectively. It is found that for the PCMO/STON diode, the V_{BI}^C is identical to the V_{BI}^J .

This agreement indicates that the PCMO/STON is a p-n diode. On the other hand, for the LSMO/STON diode, the V_{BI}^{C} deviates from the V_{BI}^{J}, because the LSMO/STON diode is a Schottky diode.

Next, they evaluate ϕ_B of the LSMO/STON Schottky diode from the temperature dependence of J_S. Based on the Schottky model, the dependence of J_S extracted from $J_F(V)$ on T follows the relation of $\ln J_S/T^2 \sim q\phi_B/k_BT$. Figure 3.16 exhibits $J_S/T^2 - 1/T$ characteristics of both diodes. Above T=150 K, the result of the LSMO/STON diode follows straight lines as predicted, and the slopes give ϕ_B of ~0.63 eV. For comparison, the same analysis of the PCMO/STON diode gives ϕ_B of ~0.70 eV.

On the basis of the above experimental results, Sawa *et al.* propose possible band diagrams of the PCMO/STON p-n diode and the LSMO/STON Schottky diode, as shown in Figs. 3.17(c) and 3.17(d), respectively. For the PCMO/STON p-n diode, because of the large dielectric constant (ε_s) and the small doping concentration in the STON depletion layer, the depletion layer width (W_D) in the STON region is much larger than that in the PCMO region. Assuming that the ε_s in the STON depletion layer is equal to that of the non-doped STO bulk (~300 at room temperature), the values of W_D in the STON region are estimated from the slopes of $1/C^2 - V$ curves as ~150 and ~110 nm for the PCMO/STON and LSMO/STON diodes at room temperature, respectively.

The high rectification has also been observed in $Nd_{0.5}Sr_{0.5}MnO_3$(NSMO)/STON heterojunctions by Matsuno *et al.* [27]. In their experiment, the (110) oriented substrates were used. That is because the (110) NSMO films can exhibit large magnetoresistance near the first-order metal-insulator transition(MIT) region which is quite different from the behavior of (001) oriented films [28]. They deposited epitaxial NSMO films (70 nm thick) on STON(110) substrates by pulsed laser deposition technique using KrF excimer laser pulses focused on NSMO ceramic targets. Laser fluence and repetition were 1-2 J /cm^2 and 5 Hz, respectively. The substrate temperature was 900 °C and the oxygen pressure was 0.4 Pa. Figure 3.18 depicts the temperature dependence of resistivity of NSMO film with and without magnetic fields. The MIT induced by temperature and magnetic field has been confirmed for the NSMO(110) thin film. The large change in resistivity with application of a magnetic field is similar to that in bulk single crystals due to the collapse of the charge ordered state [29].

With these NSMO film with MIT transition, they produced the

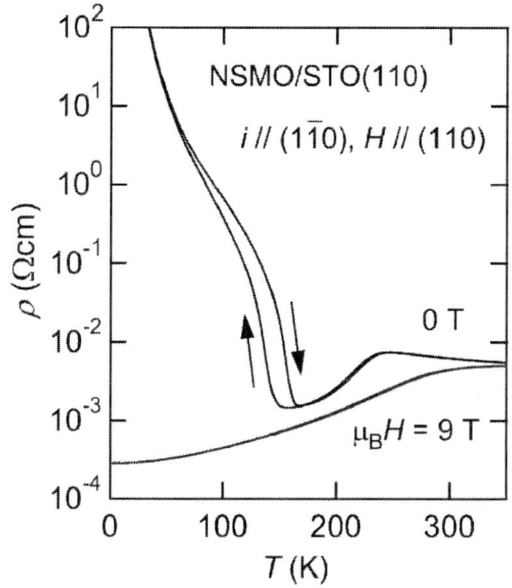

Figure 3.18. The temperature dependence of resistivity for a thin film of NSMO grown on STO(110). The solid and dashed lines represent the resistivities at 9 and 0 T, respectively. The arrows represent the hysteresis (Reproduced with permission from Reference [27]. Copyright 2008, AIP Publishing LLC).

NSMO/STON junction subsequently. To provide the NSMO films with Ohmic contact, Pt films (20 nm thick) and Au films (280 nm thick) were $exsitu$ deposited in succession. Contacts to the STON substrates were provided by the Ag paste on the backside. The Au/Pt/NSMO structure was patterned into 200×200 μm^2 mesas by conventional photolithography and Ar-ion milling. The $J-V$ characteristics were measured by the two-point contact method. The positive bias is defined as the current flow from the STON substrate to the Au electrode. Figure 3.19 shows the $J-V$ characteristics of the NSMO/ STON junctions. The junctions showed rectifying $J-V$ characteristics consistent with those of a Schottky (or a p-n) junction composed of a deep-work-function metal (or a p-type semiconductor) and an n-type semiconductor. No apparent breakdown is observed for the measured temperature range even with reverse bias voltage 100 V. According to the Schottky model, J under the forward bias is expressed as a function of V as

$$J = A^{*}T^{2}exp(-e\phi_{B}/k_{B}T)exp(eV/nk_{B}T) \tag{17}$$

where e is the electron charge, n an ideality factor, k_B the Boltzmann constant, T the temperature, A^* Richardson's constant, and ϕ_B the Schottky barrier height. This equation, which is valid for the bias region of $V \gg k_BT$, indicates that log$J$ increases linearly with V, as denoted by the broken straight lines in Fig. 3.19(b). From the slope of the log$|J|$-V line, n can be obtained. From the intersection of J at V=0, ϕ_B, which corresponds to the Fermi level position of the NSMO relative to the bottom of the conduction band of STON, can be determined. The obtained values were $n \approx$1.2 and $\phi_B \approx$0.8 eV above 100 K, which are in good agreement with those of $La_{0.7}Sr_{0.3}MnO_3$ and $Pr_{0.7}Ca_{0.3}MnO_3$. At 50 and 5 K, a large ideality factor ($n \gg 1$) was evaluated by the $|J|$-V characteristics, suggesting that the dominant contribution to the current at low temperatures is not a thermionic process but a tunneling and/or recombination process. We point out that deviation from Eq. 17 can be observed both in the higher and the lower bias regions. In the higher bias region, the current density tends to saturate due to the contribution from the series resistance of the bulk substrate. The linear relation is not realized in the lower bias region because of the tunneling effect, as shown in the case of a normal semiconductor to metal junction. The magnetic field dependence of the $|J|$-V characteristics at 5 K is displayed in Fig. 3.19(c).

3. Manganite Based Triple Layer Heterojunctions

The magnetic tunnel junctions (MTJs) are studied intensively due to their potential application in nonvolatile magnetic random access memory devices. In general, the MTJs consist two ferromagnets films which are separated by a thin insulator. If the insulating layer is thin enough (typically a few nanometers), electrons can tunnel from one ferromagnet into the other. The direction of the two magnetizations of the ferromagnetic films can be switched individually by an external magnetic field. If the magnetizations are parallel to each other, it is more likely that electrons will tunnel through the insulating film easily than if they are in the oppositional (antiparallel) orientation. Consequently, such a junction can be switched between two states of electrical resistance, one with low and one with very high resistance. The relative resistance change is defined

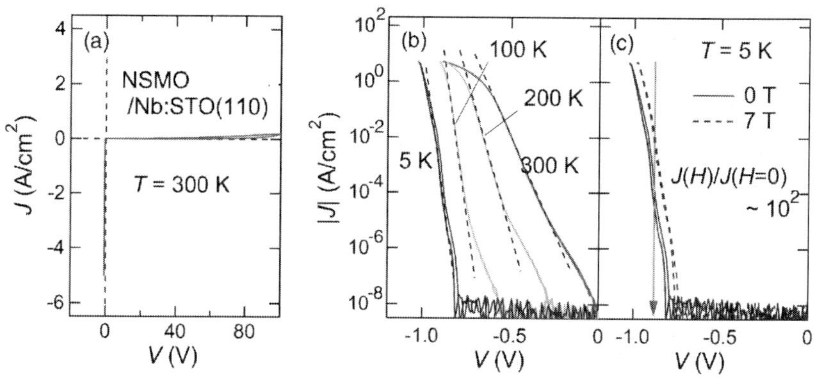

Figure 3.19. (a) $J - V$ characteristics of NSMO/STON diode measured at 300 K. (b) Temperature dependence of $|J|$-V characteristics drawn on a semilogarithmic current scale in the forward (negative) bias region. The broken straight lines indicate the linear fit with Eq.17 in the intermediate bias region. (c) Magnetic field dependence measured at 5 K. The tiny loop structures for the 5 K data represent the up and down voltage sweep (Reproduced with permission from Reference [27]. Copyright 2008, AIP Publishing LLC).

as the tunneling magnetoresistance (TMR)

$$TMR = [R_{ap} - R_p]/R_p \tag{18}$$

where R_{ap} is the electrical resistance in the antiparallel state, whereas R_p is the resistance in the parallel state. The TMR effect was explained by Jullière with the spin polarizations of the ferromagnetic electrodes [30]. The spin polarization P is calculated from the spin dependent density of states (DOS) D at the Fermi energy:

$$P = [D_\uparrow(E_F) - D_\downarrow(E_F)]/[D_\uparrow + D_\downarrow(E_F)] \tag{19}$$

The spin-up electrons are those with spin orientation parallel to the external magnetic field, whereas the spin-down electrons have anti-parallel alignment with the external field. The relative resistance change is now caused by the spin polarizations of the two ferromagnets, P_1 and P_2:

$$TMR = 2P_1P_2/[1 - P_1P_2] \tag{20}$$

If no voltage is applied to the junction, electrons tunnel in both directions with equal rates. With a bias voltage V, electrons tunnel preferentially to the positive electrode. With the assumption that spin is conserved during tunneling, the

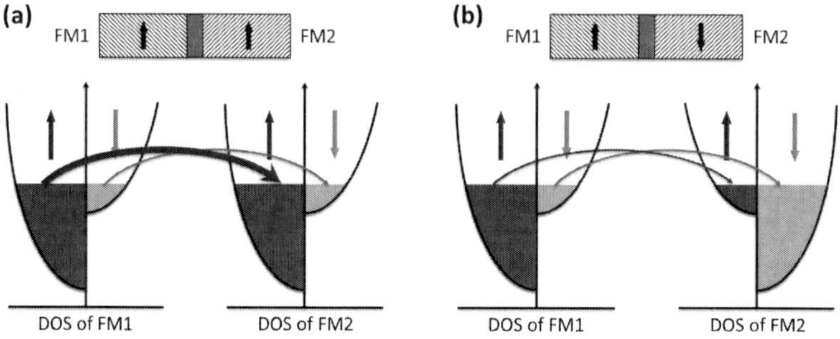

Figure 3.20. Two-current model for parallel and anti-parallel alignment of the magnetizations.

current can be described in a two-current model. The total current is split into two partial currents, one for the spin-up electrons and another for the spin-down electrons. These vary on the magnetic state of the junctions as shown in Fig. 3.20. Since this process is forbidden in classical physics, the tunnel magnetoresistance is a strict quantum mechanical phenomenon.

It is obvious that the TMR becomes infinite if P_1 and P_2 equal 1, i.e., if both electrodes have 100% spin polarization. In this case the magnetic tunnel junction becomes a switch, that switches magnetically between low resistance and infinite resistance. Such materials with 100% spin polarization are called ferromagnetic half-metals. This property is theoretically predicted for a number of materials (e.g., Fe_3O_4, CrO_2, and various Heusler alloys). Fortunately, the doped manganites of composition $La_{1-x}D_xMnO_3$ (with D=Ca, Sr, and Ba) [31, 33], and the double perovskites like Sr_2FeMoO_6 [34] or Sr_2CrWO_6 [35] are also predicted as half-metals with fully spin polarized conduction electrons. Therefore, manganites are of great interest in the fabrication of MTJs. However, it becomes clearly that the achievable TMR values are not only determined by the bulk spin polarization, as is assumed in the most simple theoretical models [30], but also sensitively depend on the electronic and structural properties of the insulating tunneling barrier as well as the insulator-ferromagnet interface.

In 1996, Sun *et al.* reported on perpendicular magnetoresistance in manganite based trilayer heterojunctions [32]. The devices were fabricated with doped perovskite manganates $La_{0.67}Ca_{0.33}MnO_3$ and $La_{0.67}Sr_{0.33}MnO_3$. Between them, a insulator $SrTiO_3$ (STO) layer with 5 nm in thickness was inserted.

It was found that large resistance changes, up to a factor of 2, can be induced by a moderate applied magnetic field below 200 Oe in these trilayers supporting current-perpendicular-to-plane transport(3.21). As they mentioned, these field-induced resistance changes originated from the spin-dependent transport across the STO barrier layer. The transport conductance across the STO barrier is related to the relative angle of the local domains inside the two LCMO layers. A change of the domain configuration in one manganite layer with respect to another will induce a change in the overall resistance of the junction. These results show that low-field spin-dependent transport in manganites can be accomplished, the magnitude of which is suitable for magnetoresistive field sensors. Unfortunately, these huge magnetoresistance can only happen at low temperatures and it is almost negligible at room temperature though the Curie temperature of LSMO is around 360 K. Recent spin polarized photoemission studies by Park *et al.* [33] demonstrated that the spin polarization in the surface layers of LSMO is 100% only at low temperature.

Mitra *et al.* reported asymmetric $I-V$ characteristics in $La_{0.7}Ca_{0.3}MnO_3/SrTiO_3/La_{0.7}Ce_{0.3}MnO_3$ trilayers [36, 37], in which the hole-doped $La_{0.7}Ca_{0.3}MnO_3$ was referred as a p-type manganite and the electron-doped $La_{0.7}Ce_{0.3}MnO_3$ presented as the n-type electrode. The inserted insulating $SrTiO_3$ layer is necessary to reduce tunneling current and avoid reactions and diffusions between the p- and n-type layers. Asymmetric $I-V$ characteristics were observed above T_C where both the p- and n-type layers displayed polaronic semiconducting behavior. At low temperatures below T_C both p- and n-type manganites behave as ferromagnetic metals and the asymmetric $I-V$ characteristics disappear.

It has been suggested that these materials should have 100% spin polarization in the ferromagnetic state and as such they could be good candidates for devices using spin polarized charge transport [38]. Except manganites, some other materials have also been suggested to have significant (close to 100%) spin polarization at room temperature. One of such materials is the half-filled ferrimagnetic oxide Fe_3O_4 [39]. To obtain significant MR similar to TMR case, Ghosh *et al.* grew high quality epitaxial heterostructures of LSMO (bottom layer) and Fe_3O_4 (top layer) with an insulating barrier layer of $SrTiO_3$ (STO) on the $LaAlO_3$ (LAO)(001) substrates and processed these for vertical transport magnetoresistance (MR) studies (Fig.3.22(a)). A very large positive magnetoresistance (MR) has been discovered in this heterostructure for the transport perpendicular to the layer planes and applied magnetic field in the film plane

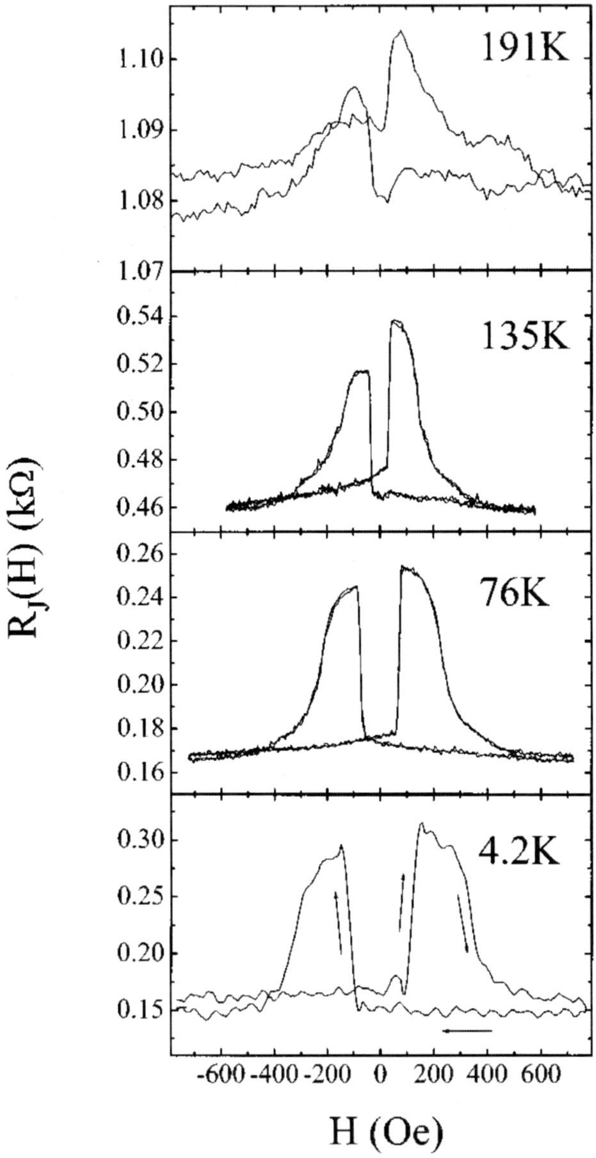

Figure 3.21. Magnetic field induced resistance modulation at various temperatures for a LSMO-based trilayer device. The pillar is a 3 μ×5.5 μm rectangle. Measurement currents were 10 μA except for the 4.2 K trace that was taken at 1 μA bias (Reproduced with permission from Reference [32]. Copyright 1996, AIP Publishing LLC).

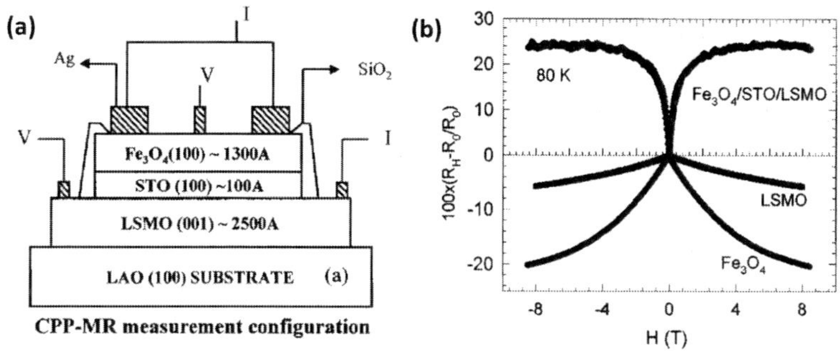

Figure 3.22. (a) Configuration of the Fe_3O_4/STO/LSMO heterostructure. (b) Magnetoresistance as a function of the in-plane applied magnetic field at 80 K for the Fe_3O_4/STO/LSMO heterostructure and for the individual layers of Fe_3O_4 and LSMO layers (Reproduced with permission from Reference [39]. Copyright 1998, AIP Publishing LLC).

(Fig.3.22(b)). The observed MR features do not show any obvious correlation with the hysteresis behavior of the ferromagnetic bilayers. They gave a possible explanation on these results that the relative differences in the majority and minority spin bands of the two ferromagnetic layers and the field induced modifications of domain structures therein. The absence of any clear correlation between the MR data and the coercivities of the ferromagnetic layers is intriguing and emphasizes the need for further exploration of interface magnetism and related phenomena. Nonetheless the effects are interesting and make new possibilities of generating novel magnetotransport phenomena in heterostructure systems.

For the properties of devices based on oxide multilayers, interfaces play a crucial role. For ferromagnetic tunnel junctions, the sign of the spin polarization depends on the barrier material and the interface termination. Several years later (in 2004), J. Simon *et al.* present a transmission electron microscopy study of heteroepitaxial trilayer systems consisting of a $SrTiO_3$ insulating barrier sandwiched between $La_{2/3}D_{1/3}MnO_3$ (D=Ca or Ba) electrodes grown on $SrTiO_3$ substrates [40]. Intermixing during the growth depends on the earth alkaline ion: Ca is segregated to the surface of the growing epitaxial thin film, leading to a Ca rich interface layer between the bottom electrode and the barrier (Fig.3.23). In contrast, no Ba segregation could be detected in the well lattice-matched Ba-doped film. In addition, a significant diffusion of both Ca

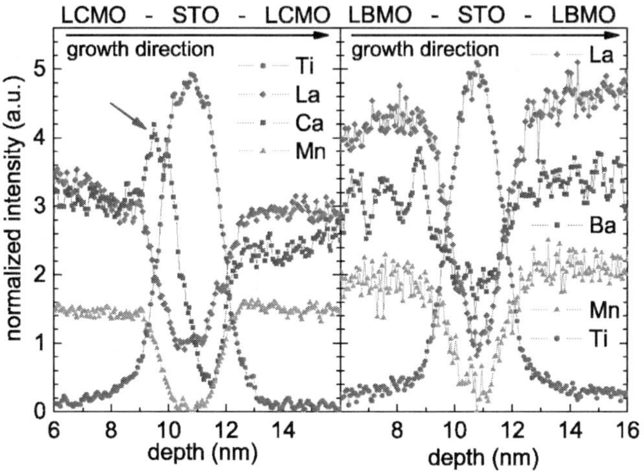

Figure 3.23. Left: Elemental profiles of Ti, La, Ca, and Mn in the LCMO/STO/LCMO trilayer structure. Right: Elemental profiles of Ti, La, Ba, and Mn in the LBMO/STO/LBMO trilayer (Reproduced with permission from Reference [40]. Copyright 1998, AIP Publishing LLC).

and Ba into the $SrTiO_3$ tunneling barrier is found, whereas the diffusion of Mn is negligibly small. Their results might be the explanation for the different tunneling magnetoresistance results observed in various manganite based tunnel junctions.

To realize the MTJs with manganite films, our group also made some efforts on this issue [41, 20]. Different from previous results, we fabricated *p-i-n* heterojunctions by depositing p-type $La_{0.67}Sr_{0.33}MnO_{3-\delta}$ (LSMO) or $Pr_{0.7}Ca_{0.3}MnO_{3-\delta}$ (PCMO) on n-type Si with a buffer layer of SiO_2. The film growth were done by the off-axis dc-magnetron sputtering technique with the substrate temperature of 750 °C. Before sputtering, the Si wafers were kept at 750 °C with oxygen pressure of 5 Pa for 5 minutes to grow SiO_2 on Si. The thickness calibrations of the SiO_2 layer (~3 nm) are carried out using ex situ spectroscopic ellipsometry. The manganite films were ~200 nm in thickness measured by field emission scanning electron microscope (FESEM).

The temperature dependence of the resistances of LSMO film at zero field and fields of 0.5, 5 T are shown in Fig. 3.24(a). The static field was applied parallel to the film plane. The metal-insulator transition at T_{MI}=208 K is observed.

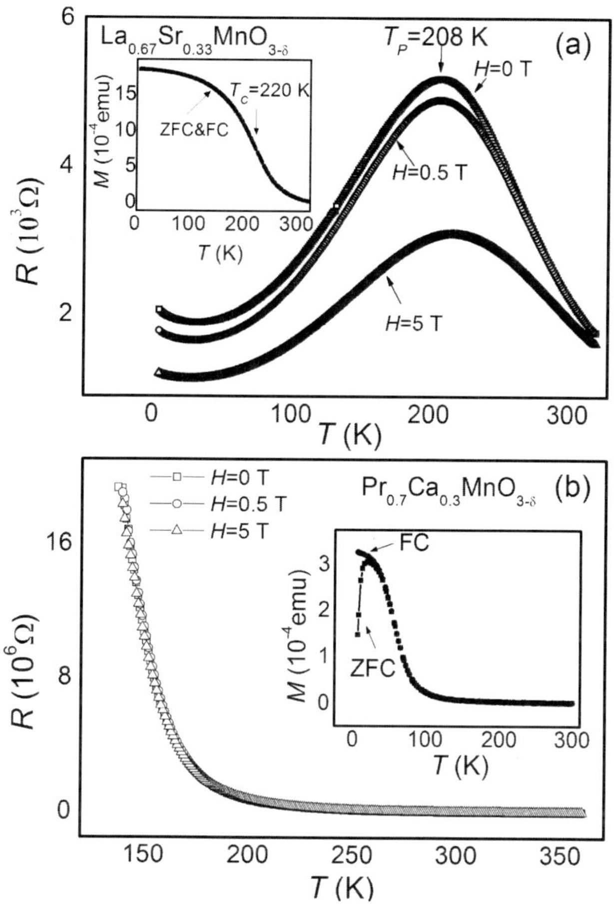

Figure 3.24. Temperature dependent resistances of LSMO (a) and PCMO (b) films in selected magnetic fields 0, 0.5, and 5 T. The inset shows temperature dependence of magnetizations M of LSMO (a) and PCMO (b) films at H=0.1 T (©IOP Publishing. Reproduced with permission from Reference [41]. All rights reserved).

In the inset of Fig. 3.24(a), the zero-filed cooled (ZFC) temperature dependence of magnetization M(T) is shown for the LSMO film. Measurements were performed with a field of 0.1 T applied within the plane of the films. The Curie temperature T_C of the LSMO films (defined as the one corresponding to the peak of dM/dT in the $Mvs.T$ curves) is about 220 K, which is much lower than that of oxygen stoichiometric film (T_C=352 K). Moreover, the temperature range of a FM-PM phase transition of LSMO film becomes broader, implying a wider distribution of the magnetic exchange interactions in the Mn-O-Mn network, i.e., the increase of magnetic inhomogeneity. In order to find the reason for the decreasing of T_C of LSMO films, the LSMO film has been annealed for 40 minutes in flowing oxygen at T=850 °C. The magnetization of the annealed sample was measured in the same way and the T_C is about 350 K (not shown here), which implies that the decreasing of T_C in oxygen-deficient LSMO films is associated with oxygen deficiency that is in agreement with the literatures. The temperature dependence of the resistances of PCMO at zero field and fields of 0.5, 5 T are shown in Fig. 3.24(b). It shows that the temperature dependent resistance curves display the semiconducting behavior ($dR/dT < 0$) in both a high-temperature PM phase and a low-temperature FM phase and the resistance maximum increases by three orders of magnitude compared with that of the LSMO sample implying the enhancement of the localization of carriers. In addition, from the inset of Fig. 3.24(b), it can be seen clearly that the ZFC curve does not coincide with the filed cooled curve below a freezing temperature T_f for the PCMO film.

The $I-V$ characteristic curves of LSMO/SiO_2/Si and PCMO/SiO_2/Si junctions measured at different temperatures ($T > T_{MI}$ for LSMO/SiO_2/Si junction) are shown in Fig. 3.25. The inset shows a schematic illustration of the current and voltage contact configuration for *I-V* measurement. The positive bias was defined as the current flows through the heterojunction from manganite films to n-type Si substrate. To minimize the heating effect at low temperature, the measuring current was kept below 1 μA. Sufficient rectifying behavior was observed in our LSMO/SiO_2/Si and PCMO/SiO_2/Si junctions. One can notice the clear asymmetry for forward and reverse biases and the current increases quickly at the diffusion voltage (V_d) under forward bias in Fig. 3.25. The current increases sharply with forward voltage as voltage exceeds diffusion potential (V_d), implying the forward conductive resistance is weak. As seen from Fig. 3.25, V_d of the LSMO/SiO_2/Si *p-i-n* junction is determined to be ~0.1 V and 0.48 V at 300 and 225 K respectively, V_d of the PCMO/SiO_2/Si *p-i-n* junction is about 0.2 V and 0.55 V at 300 and 225 K respectively, which are larger than those in LSMO/SiO_2/Si *p-i-n* junction. Moreover, the results indicate that V_d decreases with increasing temperatures, which is similar to the results in manganites/STON *p-n* junctions. It can be understood by considering a picture that depletion layer in this junction becomes thinner with increasing temperatures due to thermal diffusion, similar to the conventional *p-i-n* junctions.

In order to check if the intermediate SiO_2 layer served as the insulator layer or not, we measured the resistance of the junction ($\sim 10^6$ ohms), which was larger than manganites/STON ($\sim 10^3$ ohms). The high resistance of the junction means the key role of insulator barrier SiO_2. However, through the ratio of forward current and reverse current approaches 8 in voltage bias region from 0 to 2 V, the leakage current occurring in the reverse bias region is not very small. It reveals that there exist some pinholes in interface of these heterojunctions.

With the decreasing of temperature, the electrical behavior of LSMO films changes from insulator to metal at T_{MI}. As proved by various experiments, the manganites undergo a change of band structure accompanying paramagnetic insulator (PMI) to ferromagnetic metal (FMM) transition with a decrease of temperature, i.e., the spin-up and spin-down sub-bands split in FM state of LSMO film. So, it should be of significance to test whether the observed PMI-FMM transition in LSMO films can affect the electrical transport of LSMO/SiO_2/Si *p-i-n* junction. The experimental results are shown in Fig. 3.26. It is obvious that the electrical transport of LSMO/SiO_2/Si change a lot as the temperature decreases below T_{MI}. For the reverse bias, the current increases sharply as the reverse bias over than V_b. It should be noted that the $V_b s$ are close to each other when $T < T_{MI}$ as shown in Fig. 3.26. However, as for the forward bias, the situation becomes quite different. It is shown that the current increase slowly with the increasing of forward bias. For example, the current is 0.26 μA for bias voltage *V*=6.5 V at *T*=75 K, while the current reach 0.72 μA as *V*=-6.5 V as shown in inset of Fig. 3. The interesting asymmetry behavior is quite like that of backward diodes in the conventional semiconductor which has wide applications. The backward diode like behavior is quite different from normal diode behavior as $T > T_{MI}$.

It is well known that in connection with the tunnel diode, when the doping concentrations on the *p*- and *n*- sides of a junction are nearly or not quite degenerate, the current in the "reverse" direction for small bias, as shown in Fig. 3.26, is larger than the current in the "forward" direction, hence the name "backward diode". At thermal equilibrium, the Fermi level in the backward diode is very close to the band edges. When a small reverse bias (*p*- side is negative with respect to the *n*- side) is applied, electrons can readily tunnel from the valence band of *p*- side into conduction band of *n*- side and give rise to a tunnelling

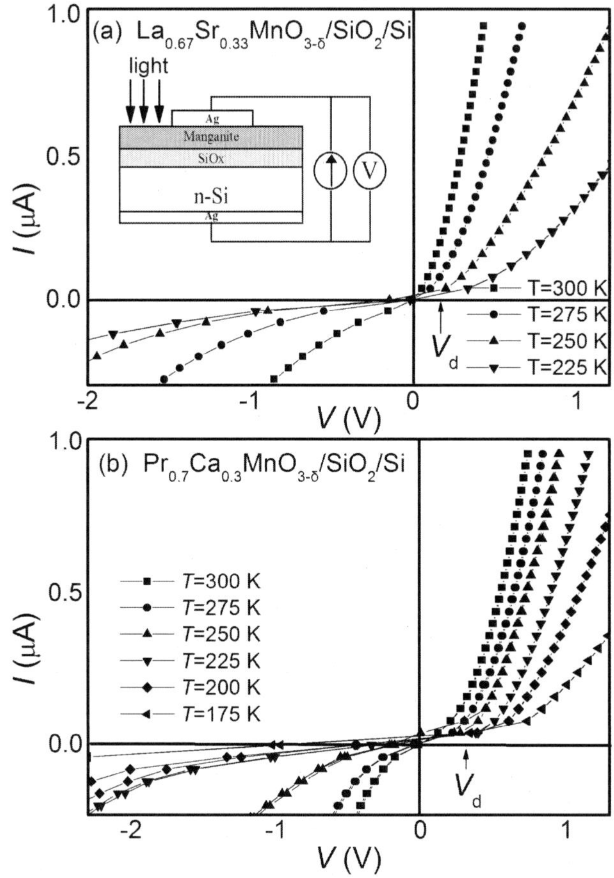

Figure 3.25. $I - V$ characteristics of the LSMO/SiO_2/Si (a) and PCMO/SiO_2/Si (b) *p-i-n* junctions measured at selected temperatures $T > T_{MI}$. The inset of (a) shows the schematic illustration for the current and voltage contact configuration for $I - V$ measurement (©IOP Publishing. Reproduced with permission from Reference [41]. All rights reserved).

current, given by the equation which can be written in the form

$$I \propto B_1 exp(+ \mid V \mid / B_2) \quad (21)$$

where the B_1 and B_2 are positive quantities and are slowly varying functions of the applied voltage *V*. Equation 21 indicates that the reverse current increases approximately exponentially with the voltage which is agreeable with our results as shown in the inset of Fig. 3.26.

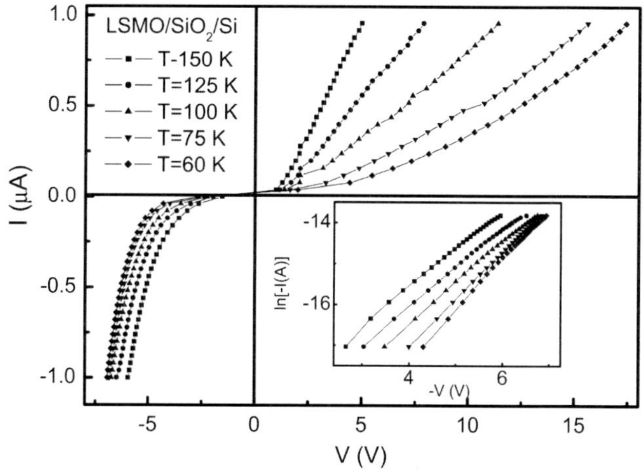

Figure 3.26. $I-V$ characteristics of the LSMO/SiO_2/Si *p-i-n* junction measured at temperatures $T < T_{MI}$. The inset shows reverse-bias ln(I)-(V) characteristics for the LSMO/SiO_2/Si *p-i-n* junctions at temperatures $T < T_{MI}$ (©IOP Publishing. Reproduced with permission from Reference [41]. All rights reserved).

According to the double-exchange model, the hopping energy of the eg electron carrier depends significantly on the relative alignment of the spins of Mn ions. At $T > T_{MI}$, the hopping energy is insufficient to overcome the charge fluctuation energy (E_{CF}) and hence the system is in the insulating phase. At $T < T_{MI}$, the spins of Mn ions start to align ferromagnetically as indicated in Fig. 3.24, enhancing the hopping energy. This increase in hopping energy may affect the small polaron energy and the Jahn-Teller distortion, and reduces E_{CF}. The reduction of E_{CF} and the increase in hopping energy lead to the band gap closing and the $n(\varepsilon_F)$ increasing upon cooling. Recently, J. H. Park *et al.* investigated the electronic structure associated with the ferromagnetic transition of $La_{1-x}Ca_xMnO_3$ and $La_{1-x}Pb_xMnO_3$ using high resolution photoemission [42]. They found that the band structure of manganites changed a lot and the density of states at the Fermi level increases as the temperature was cooling below T_{MI}. As for LSMO used in our experiments, three Mn 3d electrons form localized t_{2g} band and the remaining electrons formed eg band. As $T > T_{MI}$, the LSMO film is paramagnetic insulator and the Fermi level locates between valence band and conductor band as shown in Fig. 3.27(a). As the temperature is below T_{MI}, The eg band split into the spin-up and spin-down sub-bands and form the valence and conductor band respectively. With decreasing of temperature, the LSMO film changes from PMI to FMM accompanying the moving of Fermi level as shown in Fig. 3.27(b). That is to say the electrons are easy to occupy the low sub-band of e_g band. As proved by some experiments, the heavily doped manganite materials (LaSr)MnO_3 are degenerate semiconductor [8, 5]. Our junction is precisely the type of degenerate p^+-i-n junction below T_{MI}, and a p-i-n one above T_{MI}, considering the temperature dependence of resistance. According to this scenario, electrons can readily tunnel from the valence band of p^+- side into the conduction band of n- side and give rise to a tunnelling current when a small reverse bias (p^+- side negative with respect to the n- side) is applied. Therefore, the backward behavior emerges and this feature has been proved by our experiments. The backward diode can be used for rectification of small signals, microwave detection, and mixing. It is expected that the emergence of backward property may exert potential application of manganite-based heterojunctions.

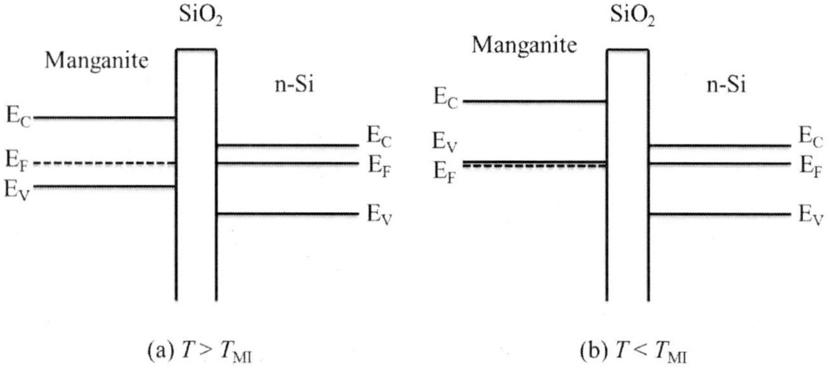

Figure 3.27. A possible energy-band diagram for LSMO/SiO_2/Si p-i-n junction with temperature $T > T_{MI}$ (a) and $T < T_{MI}$ (b) (©IOP Publishing. Reproduced with permission from Reference [41]. All rights reserved).

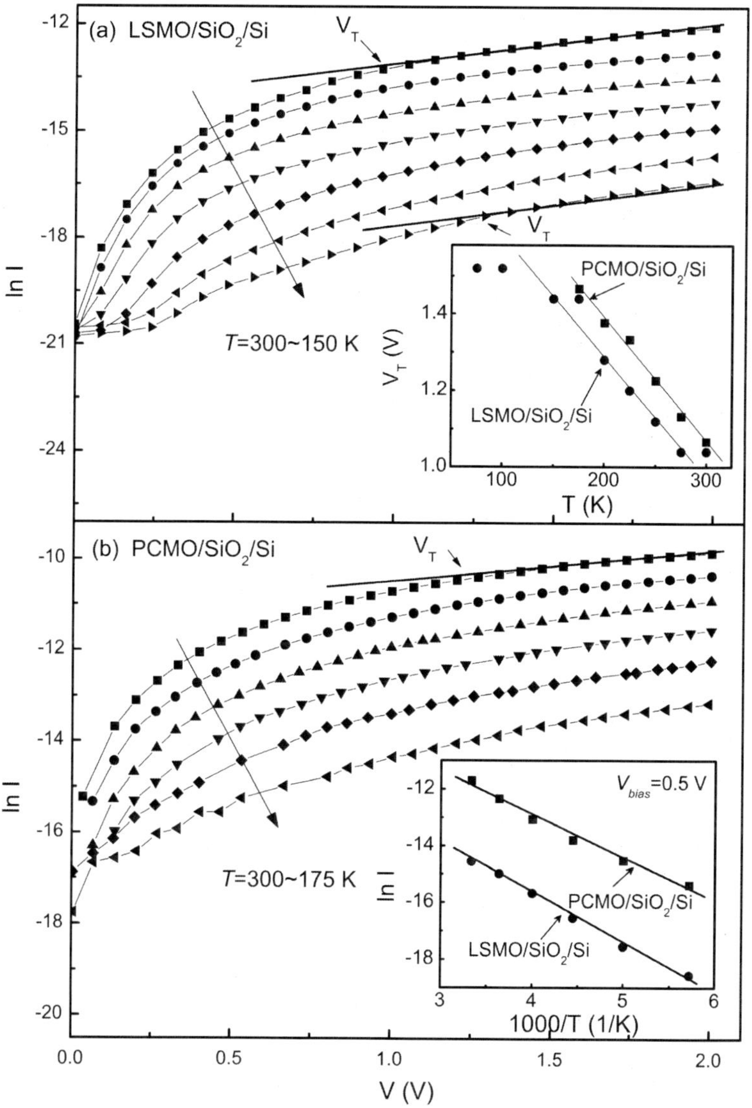

Figure 3.28. Forward bias $I - V$ characteristics for the LSMO/SiO_2/Si (a) and PCMO/SiO_2/Si (b) *p-i-n* junctions. The inset of (a) shows temperature dependent V_T for the LSMO/SiO_2/Si and PCMO/SiO_2/Si *p-i-n* junctions. The inset of (b) shows the current versus temperature relationship at a bias voltage of 0.5 V. The straight lines are a guide for the eyes (©IOP Publishing. Reproduced with permission from Reference [41]. All rights reserved).

Because of the existence of the insulator barrier SiO_2, the *p-i-n* heterojunction we studied here is one kind of tunnel junctions. In order to explore the transport mechanism of this kind of junction further, the current-voltage characteristics have been paid more attention to. Figure 3.28(a) shows the $I - V$ curves with a positive voltage bias of the LSMO/SiO_2/Si *p-i-n* junction. The current is on a logarithmic scale. Several features of these $I - V$ characteristics are noteworthy.

First, the current depends exponentially on the voltage as the bias voltage exceeds a threshold value. We defined the threshold voltage where the ln(I)-V relationship becomes linear as V_T. Moreover, the curves as $V > V_T$ are parallel as shown in figure 3.28, indicating that the slopes of the $I - V$ curves do not depend on temperature. This behaviour is consistent with the tunnelling current mechanism in that the current can be expressed as

$$I \sim exp(\beta V) \tag{22}$$

where β is a temperature independent coefficient and V is the applied voltage. This process can be understood as follows. With the forward bias voltage applied, the potential of n-Si is enhanced and the electrons in the conduction band of n-Si can tunnel to a corresponding state in manganite films directly once the voltage is larger than V_T. The inset of figure 5(a) shows the temperature dependence of V_T for both LSMO/SiO_2/Si and PCMO/SiO_2/Si *p-i-n* junctions. It reveals that the tunnel threshold voltage for the PCMO/SiO_2/Si junction is larger than that for the LSMO/SiO_2/Si junction. It can also be found that the V_T is linearly increasing with decreasing temperature for the PCMO/SiO_2/Si *p-i-n* heterojunctions in the whole measurement temperature range. As for the LSMO/SiO_2/Si junction, the relation of $V_T - T$ is approximately linear as $T >$150 K. When the temperature is below 150 K, the V_T does not change with the decrease in temperature. This feature indicates the difference in the electrical transport mechanism between lower and higher temperatures in the LSMO/SiO_2/Si junction once again. This feature agrees with the above analysis about the change in the electronic band structure of the LSMO/SiO_2/Si junction as temperature decreases.

The second feature in $I - V$ characteristics we should notice is that the relationship of ln(I)-V is not linear at a lower bias voltage ($V < V_T$) as shown in Fig. 3.28. In this case, the electrical structures of the i-SiO_2 layer should be considered. It is well known that there is a well-defined energy gap between the conduction and valence bands (about 9 eV at 300 K) in a crystalline

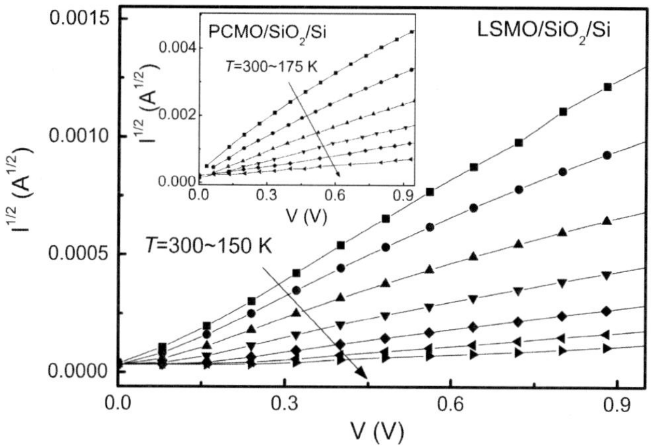

Figure 3.29. The squared currents as functions of bias voltage ($I^{1/2} - V$) at lower voltages of manganite/SiO_2/Si *p-i-n* junctions (©IOP Publishing. Reproduced with permission from Reference [41]. All rights reserved).

SiO_2 solid. However, in an amorphous SiO_2 layer, the lack of long-range order in a non-crystalline solid causes smearing of the conduction and valence band edges, so that the energy gap is no longer well defined [14]. Because the band edges are diffused, there is a gradual transition from quasi-continuous states (conduction and valence bands), in which the carriers can move freely to strictly localized states (density of states tail) or traps in which the carriers are immobilized. Moreover, a number of trap centres in the amorphous SiO_2 layer would cause the so-called space-charge-limited effect [16]. In the space-charge-limited transport process, the carriers (electrons) will be injected into the conduction band of the i-SiO_2 layer from the n layer and flow towards the *p* layer together with the residual carriers if the i-SiO_2 layer involves some residual conduction carriers with a forward bias voltage. The space-charge-limited current is expressed by

$$I \sim (V^2/N_t)exp(-E_t/k_BT) \tag{23}$$

where N_t and E_t are the trap density and the trap level, respectively, in the i-SiO_2 layer. *V* is the voltage consumed in the i-SiO_2 layer to overcome the extra barrier caused by the trapped space charge. From equation 23, it can be seen that the square root current $I^{1/2}$ will be linear with the bias voltage *V* at a fixed temperature if the space-charge-limited transport is dominant in the junction. This feature can be found in Fig. 3.29. It shows that the curves are linear, i.e., the current increases with V^2 at a low voltage bias which is consistent with the space-charge-limited mechanism. In contrast to the tunnelling process, the space-charge-limited current is very sensitive to the measurement temperature T, and the logarithmic space-charge-limited current will change linearly with the reciprocal T at a constant voltage. In the inset of Fig. 3.28(b), the current values at 0.5 V measured at various temperatures in Fig. 3.29 are replotted in terms of the $\ln(I)\sim(1000/T)$ relation, where data fall on straight lines, suggesting that the observed temperature dependence of the $I-V$ characteristics is dominated by the space-charge-limited rule at $V < V_T$. The trap level, calculated from the slope of the lines based on equation 23, is 0.15 eV for LSMO/SiO_2/Si and 0.13 eV for PCMO/SiO_2/Si junctions, respectively, being quite deep. From the above analysis, one can possibly arrive at the conclusion that the space-charge-limited effect may dominate at a low voltage bias ($V < V_T$) and the tunnelling mechanism may be dominant when $V > V_T$ in our manganite/SiO_2/Si *p-i-n* junctions. Therefore, it is easy to understand that the slopes of the $I-V$ curves show strong temperature dependence as $V < V_T$ and are temperature independent when $V > V_T$ as shown in figure 3.28.

References

[1] M. B. Salamon, M. Jaime, *Rev. Mod. Phys.* 73, 583(2001).

[2] C. Mitra, P. Raychaudhuri, K. Dörr, K.H. Müller, L. Schultz, P.M. Oppeneer, S. Wirth, *Phys. Rev. Lett.* 90, 017202(2003).

[3] S. W. Han, J. -S. Kang, K. H. Kim, J. D. Lee, J. H. Kim, S. C. Wi, C. Mitra, P. Raychaudhuri, S. Wirth, K. J. Kim, B. S. Kim, J. I. Jeong, S. K. Kwon, B. I. Min, *Phys. Rev., B* 69, 104406(2004).

[4] T. Yanagida, T. Kanki, B. Vilquin, H. Tanaka, T. Kawai, *Phys. Rev. B* 70, 184437(2004).

[5] H. Tanaka, J. Zhang, T. Kawai, *Phys. Rev. Lett.* 88, 027204(2002).

[6] N. Nakagawa, M. Asai, Ymukonoki, T. Susaki, H. Y. Hwang, *Appl. Phys. Lett.* 86 (2005) 082504.

[7] J. R. Sun, S. Y. Zhang, B. G. Shen, H. K. Wong, *Appl. Phys. Lett.* 86, 053503(2005).

[8] T. Kudo, M. Tachiki, T. Kashiwai, and T. Kobayashi, *Jpn. J. Appl. Phys. Part 2* 69, L999 (1998).

[9] J. Zhang, H. Tanaka, and T. Kawai, *Appl. Phys. Lett.* 80, 4378 (2002).

[10] J. R. Sun, C. M. Xiong, T. Y. Zhao, S. Y. Zhang, Y. F. Chen, B. G. Shen, *Appl. Phys. Lett.* 84, 1528(2004).

[11] A. Sawa, T. Fujii, M. Kawasaki, and Y. Tokura, *Appl. Phys. Lett.* 86, 112508(2005).

[12] Kui-juan Jin,Hui-bin Lu, Qing-li Zhou, Kun Zhao, Bo-lin Cheng, Zheng-hao Chen, Yue-liang Zhou, and Guo-Zhen Yang, *Phys. Rev. B* 71, 184428(2005).

[13] T. Susaki, N. Nakagawa, and H. Y. Hwang, *Phys. Rev. B* 75, 104409(2007).

[14] S. M. Sze, Physics of Semicondutor Devices, 2nd ed, Wiley, New York, 1981.

[15] Z. Sroubek, *Phys. Rev. B* 2, 3170 (1970).

[16] M. Sugiura, K. Uragou, M. Tachiki, and T. Kobayashi, *J. Appl. Phys.* 90, 187 (2001).

[17] J. R. Gebhardt, S. Roy, N. Ali, *J. Appl. Phys.* 85, 5390(1999).

[18] C. Mitra, P. Roychaudhuri, J. John, S.K. Dhar, A.K. Nigam, R. Pinto, *J. Appl. Phys.* 89, 524(2001).

[19] J. R. Sun, C. M. Li, H. K. Wong, *Appl. Phys. Lett.* 84, 4804(2004).

[20] Z. G. Sheng, W. H. Song, Y. P. Sun, J. R. Sun, B. G. Shen, *Solid State Communications* 137, 292-295(2006).

[21] T. Shimizu and H. Okushi, *J. Appl. Phys.* 85, 7244(1999).

[22] Y. Tomioka, A. Asamitsum, H. Kuwahara, Y. Moritomo, and Y. Tokura, *Phys. Rev. B* 53, R1689 (1996).

[23] Y. Tomioka and Y. Tokura, *Phys. Rev. B* **70**, 014432 (2004).

[24] P. T. Landsberg, *Proc. Phys. Soc. London, Sect. B* 65, 397 (1952).

[25] R. Cooper, *Proc. Phys. Soc. London, Sect. B* 65, 409 (1952).

[26] T. Fujii, M. Kawasaki, A. Sawa, H. Akoh, Y. Kawazoe, and Y. Tokura, *Appl. Phys. Lett.* 86, 012107 (2005).

[27] J. Matsuno, A. Sawa, M. Kawasaki, and Y. Tokura, *Appl. Phys. Lett.* 92, 122104(2008).

[28] Y. Ogimoto, M. Nakamura, N. Takubo, H. Tamaru, M. Izumi, and K. Miyano, *Phys. Rev. B* 71, 060403(R) (2005).

[29] H. Kuwahara, Y. Tomioka, A. Asamitsu, Y. Moritomo, and Y. Tokura, *Science* 270, 961 (1995).

[30] M. Jullière, "Tunneling between ferromagnetic films". *Phys. Lett.* 54A, 225-226(1975).

[31] R. von Helmholt, J. Wecker, B. Holzapfel, L. Schultz, and K. Samwer, *Phys. Rev. Lett.* 71, 2331(1993).

[32] J. Z. Sun, W. J. Gallagher, P. R. Duncombe, L. Krusin-Elbaum, R. A. Altman, A. Gupta, Y. Lu, G. Q. Gong, and Gang, *Appl. Phys. Lett.* 69, 3266(1996).

[33] J. H. Park, E. Vesoovo, H. J. Kim, C. Kwon, R. Ramesh, and T. Venkatesan, *Nature (London)* 392, 794 (1998).

[34] M. Sugiura, K. Uragou, M. Noda, M. Tachiki, T. Kobayashi, *Jpn. J. Appl. Phys. Part 1*, 38, 2675(1999).

[35] J. B. Philipp, D. Reisinger, M. Schonecke, A. Marx, A. Erb, L. Alff, R. Gross, and J. Klein, *Appl. Phys. Lett.* 79, 3654(2001).

[36] C. Mitra, P. Raychaudhuri, G. Kobernik, K. Dorr, K.-H. Muller, L. Schultz, and R. Pinto, *Appl. Phys. Lett.* 79, 2408(2001).

[37] C. Mitra, G. Kobernik, K. Dorr, K.-H. Muller, L. Schultz, P. Raychaudhuri, R. Pinto, and E. Wieser, *J. Appl. Phys.* 91, 7715(2002).

[38] H. Y. Hwang, S. W. Cheong, N. P. Ong, and B. Batlog, *Phys. Rev. Lett.* 77, 2041 (1996).

[39] K. Ghosh, S. B. Ogale, S. P. Pai, M. Robson, Eric Li, I. Jin, Zi-wen Dong, R. L. Greene,R. Ramesh, T. Venkatesan, and M. Johnson, *Appl. Phys. Lett.* 73, 689(1998).

[40] J. Simon, T. Walther, W. Mader, J. Klein, D. Reisinger, L. Alff, and R. Gross, *Appl. Phys. Lett.* 84, 3882(2004).

[41] Z. G. Sheng, Y. P. Sun, X. B. Zhu, J. M. Dai, W. H. Song, Z. R. Yang, L. Hu, and R. R. Zhang, *J. Phys. D: Appl. Phys.* 41, 135008(2008)

[42] J. H. Park, C. T. Chen, S. W. Cheong, W. Bao, G. Meigs, V. Chakarian, and Y. U. Idzerda, *Phys. Rev. Lett.* 76, 4215 (1996).

In: Manganite-Based Thin Films ...
Editors: Z. Sheng, L. Hu and Y. Sun ISBN: 978-1-63482-452-1

Chapter 4

PHOTOINDUCED AND PHOTOVOLTAIC EFFECT IN MANGANITE-BASED FILMS AND HETEROJUNCTIONS

***L. Hu*[1] *and Y. P. Sun*[1,2,*]**
[1] Key Laboratory of Materials Physics,
Institute of Solid State Physics, Chinese Academy of Sciences,
Hefei, People's Republic of China
[2] High Magnetic Field Laboratory, Chinese Academy of Sciences,
Hefei, People's Republic of China

Abstract

The manganites are sensitive to external stimuli (magnetic field, light, and electric field). Among them, the photoinduced effects are not only important to help understanding many intriguing physical properties observed in these materials, but also especially vital for the possible applications to data recording technology and photoelectronic devices. Cooperative interactions among microscopic degrees of freedom make it possible to switch electronic, magnetic or structural properties by light irradiation. In this chapter, the study is focused on the photoinduced and photovoltaic effects in manganite-based films and heterojunctions. Photoconductivity, PIPT, and optical-magnetic effect in manganite thin films are reviewed in the chapter. And the photovolatic effects and the effect of light irradiation on the physical properties of manganite-based heterojunction are also included.

*E-mail address: ypsun@issp.ac.cn

PACS: 75.47.Lx, 73.50.Pz

Keywords: manganite, photoinduced effects, photovoltaic effects

1. Introduction

The consequences of interaction between light and matter are significant methods to investigate the matter's physical properties. Optical absorption occurs due to the interaction of photon, electron, exciton, lattice vibration (phonon), and defect in materials. The generation of luminescence can be produced by external stumuli, such as electromagnetic field. The information of electronic structure can be obtained through investigating the optical absorption and luminescence. The manganites exhibit various kinds of electronic phases such as charge- and orbital-ordered insulator (COOI), ferromagnetic metal (FMM), ferromagnetic insulator (FMI), and paramagnetic insulator (PMI). These competing phases originate from the strongly correlated charge, spin, orbital, and lattice degrees of freedom and produce complex phase diagrams. Therefore, the manganites are sensitive to external stimuli (magnetic field, light, and electric field). Among them, the photoinduced effects, such as photoconductivity and photoinduced insulator-to-metal transition (IMT), are not only important to help understanding many intriguing physical properties observed in these materials, but also especially important in the possible applications to data recording technology and photoelectronic devices.

Saitoh *et al.* have investigated the electronic structure of $La_{1-x}Sr_xMnO_3$ by photoemission and O 1s x-ray-absorption spectroscopy. Spectra of the Mn 2p core levels and the valence bands for $LaMnO_3$ and $SrMnO_3$ have been analyzed using a configuration-interaction cluster model. The ground state of $LaMnO_3$ is found to be mixed d^4 and d^5 L states and that of $SrMnO_3$ to be heavily mixed d^3 and d^4 L states, reflecting their strong covalency. The character of the band gap of $LaMnO_3$ is of the p-to-d charge-transfer type while that of $SrMnO_3$ has considerable p-p character as well as p-d character. Holes doped into $LaMnO_3$ mainly of oxygen p character are coupled antiferromagnetically with the d^4 local moments of the Mn^{3+} ions and become itinerant, which therefore align the Mn moments ferromagnetically. A schematic picture of the spectral weight distribution and its evolution with x are shown in Fig. 4.1. The unoccupied states of hole-doped $LaMnO_3$, $e_{g\uparrow}$-like spectral weight will appear within the band gap of $LaMnO_3$. In the composition range of FMM at low temperature, the e_g

band is fully polarized due to Hund's rule coupling with the fully polarized $t_{2g\uparrow}$, so that a spinless model becomes relevant to the $e_{g\uparrow}$ electrons [1]. Okimoto *et al.* have investigated the optical conductivity spectra and their variation with temperature and doping level x for single crystal of $La_{1-x}Sr_xMnO_3$ ($0 \leqslant x \leqslant 0.3$) (Fig. 4.2). For the low-doped insulating crystal (x=0.1), the spectral weight of the optical conductivity increases only in the inner-gap region around 0.5 eV, but no Drude part emerges due to the carrier localization effect. For $x \geqslant 0.17$, where the low-temperature ferromagnetic metallic state shows up, the optical conductivity spectrum above T_C is characterized by interband transitions between the exchange-split conduction bands, and it gradually changes into that of intraband excitations below T_C . In the metallic phase, low-energy spectra arising from intraband excitations can be sorted into two parts: One is a nearly ω-independent broad structure (incoherent part), and the other a sharp coherent Drude peak with anomalously low spectral weight. This can hardly be reconciled with the simple double-exchange theory, but indicates that another degree of freedom (e.g., the orbital ordering and/or electron-lattice interactions) should be taken into account [2]. Shi *et al.* have presented angle-resolved photoemission spectroscopy (ARPES) results for thin films of $La_{1-x}Sr_xMnO_3$ (Fig. 4.3). It is found that the T_C from the PMI to FMM state is closely related to details of the electronic structure, particularly to the spectral weight at the k-point, where the sharpest step at the Fermi level is observed. The k-point is the same for all the samples, despite their different values of T_C. The change of T_C is discussed in terms of kinetic energy optimization. The ARPES results suggest that the change of the electronic structure for the samples having different transition temperatures is different from the rigid band shift [3].

The photoinduced effects in the doped manganites may lead to new strategies for the realization of all-optical high-speed switches and data storage devices. Therefore, it is essential to understand the ultrafast mechnisms. Rini *et al.* have used time-resolved x-ray absorption near-edge structure spectroscopy (XANES) to investigate the electronic dynamics associated with the photoinduced insulator-to-metal phase transition in the CMR manganite $Pr_{0.7}Ca_{0.3}MnO_3$. Absorption changes at the 0 K and Mn L edges directly monitor the evolution of the density of unoccupied states in the transient photoinduced phase and clearly indicate that this phase is metallic (Fig. 4.4). The electronic structure of the photoinduced phase is remarkably similar to that of the FMM phase reached in related manganites upon cooling below the T_C [4].

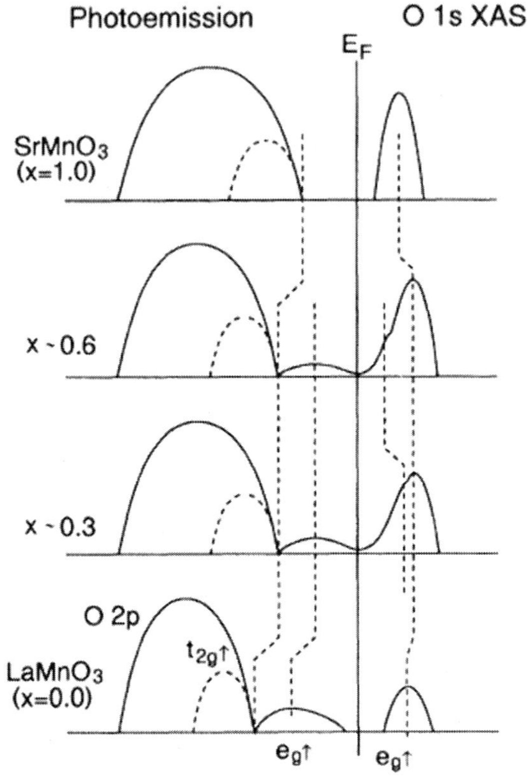

Figure 4.1. Schematic picture for the evolution of the electronic structure of $La_{1-x}Sr_xMnO_3$(Reprinted with permission from Reference [1], Copyright (1995) by the American Physical Society).

2. Photoinduced Effects in Manganite Thin Films

When the light is incident upon a material surface, the electrons present in the valence band absorb energy and, being excited, jump to the conduction band and become free.There are three important effects produced by the optical absorption: photoelectric effect, photoconductivity effect, and photovoltaic effect. They play an important role in the optelectric technology and devices. In the photoelectric effect, electrons are ejected from a material's surface into vacuum. Electrons emitted in this manner may be called photoelectrons.If the

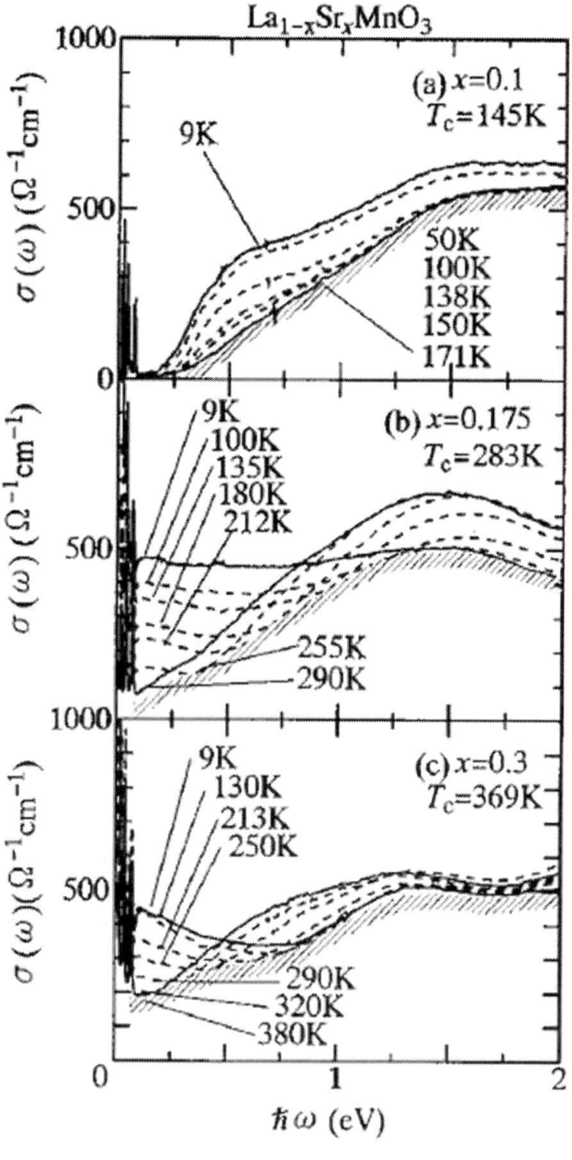

Figure 4.2. Optical conductivity spectra at various temperatures in $La_{1-x}Sr_xMnO_3$: (a) x=0.1, (b) x=0.175, and (c) x=0.3(Reprinted with permission from Reference [2], Copyright (1997) by the American Physical Society).

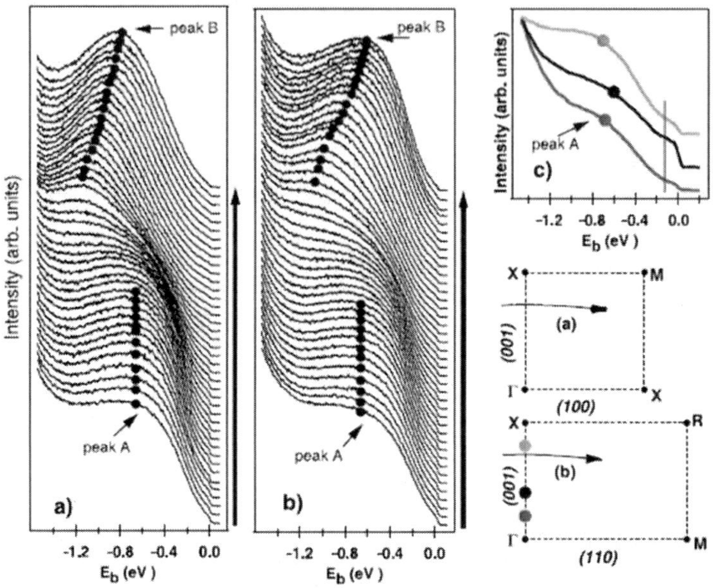

Figure 4.3. ARPES spectra for $La_{1-x}Sr_xMnO_3$ thin film (T_C =313 K) at 30 K. (100) and (110) are along the sample surface, while (001) is the surface normal(©IOP Publishing. Reproduced with permission from Reference [3]. All rights reserved).

excited electrons do not escape the surface and remain in the material, those electrons would result in the change of conductivity (photoconductivity effect) or creation of voltage/electric current (photovoltaic effect). The photoconductivity can be divided into infrared photoconductivity, visible photoconductivity, ultraviolet photoconductivity, and x-ray photoconductivity in view of the frequency of light. According to the temporal behavior, the photoconductivity can also be divided into transient photoconductivity and persistent photoconductivity (PPC).

The band gap of the doped manganites is about 1 eV. Furthermore, the doped manganites posses various electronic phases with similar free energy. Therefore, the doped manganites are susceptible to the irradiation of light. In 1997, Kiryuhkin *et al.* have shown that a transition from COOI to FMM can be driven by illumination with x-ray at low temperature (<40 K) in $Pr_{0.7}Ca_{0.3}MnO_3$. This transition is accompanied by significant changes in the lattice structure, and

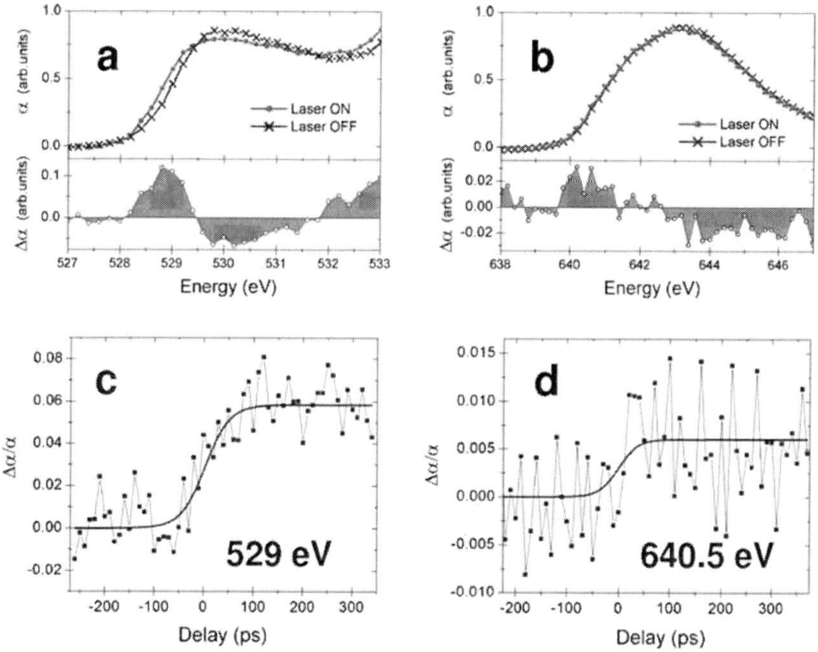

Figure 4.4. Time-resolved XANES measurements at O K and Mn L edges. Static absorption spectra (solid circles) at the (a) O K pre-edge and (b) Mn L_3 edge and spectra measured 300 ps after 800 nm excitation (crosses). (c) and (d) Relative change in absorption ($\Delta\alpha/\alpha$) as a function of pulse delay at two representative wavelengths, 529 (O K edge) and 640.5eV (Mn L edge)(Reprinted with permission from Reference [4], Copyright (2009) by the American Physical Society).

can be reversed by thermal cycling. The intensity of the reflection decreases continuously with x-ray illumination, no change is observed when the x-ray is switched off. The diminution of the charge order is associated with a dramatic change in transport properties. The conductance of the sample increases by more than six orders of magnitude after $\sim$20 min of total x-ray exposure. When the x-ray beam is switched off, the conductance persists for many hour without measurable degradation. The photoinduced transition and increased conductivity are caused by x-ray photoelectrons and secondary electrons generated in collisions [5]. Besides x-ray, the infrared and visible light can also produce obvious photoconductivity effect [6, 7]. It is clear that the light-induced insulator-metal transition is triggered by photocarrier generation in $Pr_{0.7}Ca_{0.3}MnO_3$, in which the metallic state is the result of the collapse of the charge-ordered state [6]. Fiebig *et al.* have shown that the light-induced insulator-metal transition in $Pr_{0.7}Ca_{0.3}MnO_3$ is to generate a well-localized conducting path while the bulk of the sample remain insulating. The path can be visualized through a change of reflectivity that accompanies the phase transition. The insulator-metal transition is induced by illuminating the gap between the electrodes with 7-ns, 1064-nm laser pulses while applying a voltage of typically 10 to 100 V. The phase transition can only be induced if the laser spot fully covers the space between the electrodes while the laser-pulse energy exceeds a threshold value that is determined by the electric field between the electrodes and the laser wavelength. Once the phase transition occurs , it remains stable as long as a current I flows between the electrodes, in which case the laser was turned off. The change of reflectivity ΔR due to the flowing current was visualized by digitally subtracting the image of the conducting sample from the image of the insulating sample. For I=0 mA, the sample remains in the insulating state and the difference image should not exhibit a variation of brightness at all. At I=10 mA, the width of the path is about 30 μm. At lower current, it is even narrower and may actually become filamentary, but this is not resolved by the optical system. Upon an increase to I=100 mA, the width of the path approaches 250 μm, and it exceeds 500 μm when I is further increased [7]. However, light illumination can do more than photocarrier injection in the doped manganites. Solids with strongly correlated electrons, in which even subtle crystallographic distortions can result in colossal changes of the electronic and magnetic properties, could be directed between competing phases by coherent excitation of specific infrared-active modes. Rini *et al.* have reported the ultrafast switching of the electronic phase of $Pr_{0.7}Ca_{0.3}MnO_3$ via direct excitation of a phonon mode at 71 meV (Mn-O stretching mode). This vibrational mode affects the geometrical parameters determining the tolerance factor and thus has a strong coupling to the electronic properties through a coherent modulation of the electron hopping probability. A prompt, five-order-of-magnitude drop in resistivity is observed, associated with a non-equilibrium transition from the stable insulating phase to a metastable metallic phase (Fig. 4.5) [8]. Most of the photoinduced effects in manganite single crystals are transient due to the local photoinduced insulator-metal transition. Since the penetration depth of light is typically only 0.1 μm, the photoinduced transformation is limited to the surface of the crystal, leaving

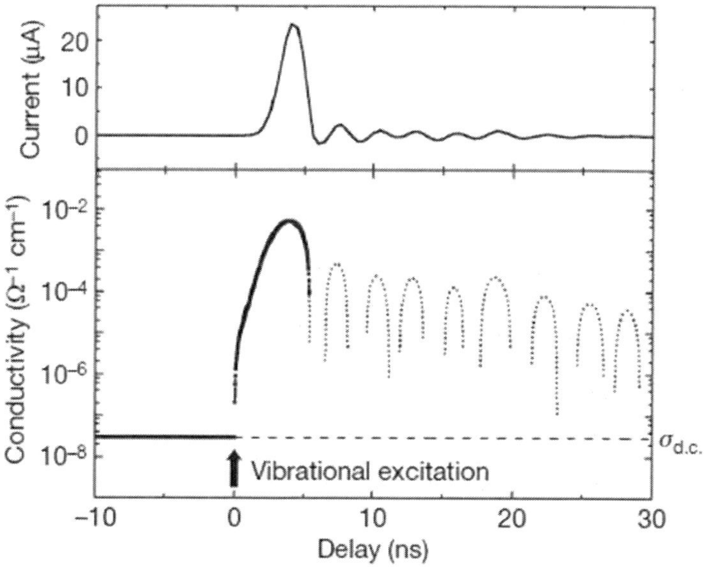

Figure 4.5. Time-dependent transport measurement. Vibrational excitation of the MnCO stretching mode results in a ,10^3 increase in the sample current (upper panel) and a corresponding ,10^5 increase in the sample conductivity (lower panel)(Reprinted by permission from Macmillan Publishers Ltd: Nature (Ref.[8]), copyright (2007)).

the bulk of the sample in the insulating state. Therefore, it has been suspected that the "pressure" from the surrounding region forces the metallic phase back into the insulator. Therefore, it is expected that the photoinduced changes can be extended to the whole sample and be persistent in manganite thin film.

Oshima *et al.* have observed the PPC effect in a Cr-doped $Pr_{0.5}Ca_{0.5}MnO_3$ thin film. The film shows a diffuse phase transition and a phase separation that arise from the quenched random fields produced by the Cr ions. Photoexcitation of the sample with a continuous-wave-yttrium-aluminum-garnet laser causes a gradual increase of the conductance in the FMM state, whereas the magnetization and absorption spectra show little change. The persistent conduction is ascribed to the increase in the conducting paths by the optical excitation in the percolative conduction through FM regions, without significant overall growth of the FM regions (Fig. 4.6) [9]. Dai *et al.* have studied the photoinduced

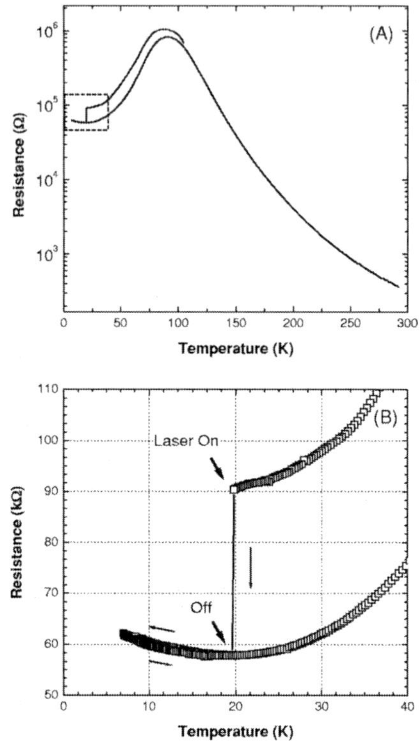

Figure 4.6. (a) Temperature dependence of the resistance before and after the photoexcitation at 20 K. (b) The behavior around 20 K presented by the dashed square in (a) is magnified(Reprinted with permission from Reference [9], Copyright (2001) by the American Physical Society).

effects in perovskite manganite thin films of $(La_{0.3}Nd_{0.7})_{2/3}Ca_{1/3}MnO_3$. The PPC effect induced by a He-Ne laser with a photon energy of $\sim$2 eV and the normal photoconductivity effect induced by a mercury lamp with photon energy near 3 eV are observed at low temperature where the system is in (or close to) the cluster-glass (CG) state (Fig. 4.7). These effects can coexist and are independent of each other. Their origins can be explained in terms of the CG state formed by the FM-metallic clusters with different orientations of magnetic domains frozen in a disorderly fashion and embedded into the matrix with the CO phase as the temperature decreases below 50 K. The PPC effect originates from photon-induced charge transfer in FM domains resulting in FM cluster growth,

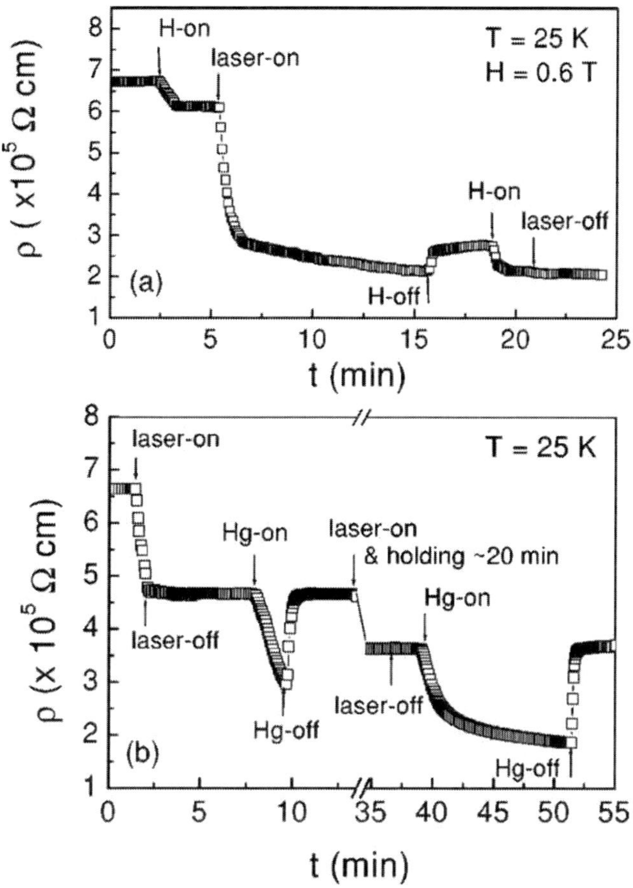

Figure 4.7. Time dependence of the resistivity on (a) the combined laser illumination and a low magnetic field of 0.6 T and (b) the combined laser and Hg-lamp illumination at 25 K(Reprinted with permission from Reference [11], Copyright (2003) by the American Physical Society).

while the normal photoconductivity effect originates from photon-induced ionization of localized carriers in the CO phase [10, 11].

As previous studies pointed out, phase competition is crucial to phase control by light, among which the photoexcitation of the electronic states causes transformation among metastable ground states [12]. Such a phase competition is prevalent in the perovskite manganites and $Pr_{0.55}(Ca_{1-y}Sr_y)_{0.45}MnO_3$

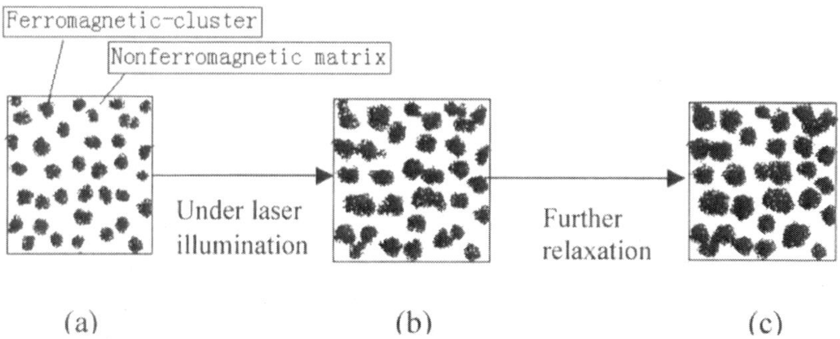

Figure 4.8. Schematic diagrams illustrate CG state and photoninducing FM cluster growth model. (a) The system in the CG state before laser illumination; (b) FM clusters growing under laser illumination; (c) the boundaries of FM clusters in CG state modified and further relaxed after laser illumination(Reprinted with permission from Reference [11], Copyright (2003) by the American Physical Society).

is a typical example of phase competition among different electronic states. For y≤0.2, the ground state is a COOI, while for y≥0.25, it is a FMM. At around y=0.25, there is a multicritical point, in which a CE-type AFM, a FMM, and a PMI compete. Takubo *et al.* have achieved persistent and reversible optical control of FMM and COOI phases in a manganite thin film of $Pr_{0.55}(Ca_{0.75}Sr_{0.25})_{0.45}MnO_3$. Pulsed laser light brings the lower temperature FMM phase out of the higher temperature COOI phase (Fig. 4.9)[13]. In the warming process of around 80 K, a persistent photoinduced metal-to-insulator transition has been observed. Together with the previously reported reverse effect, the photoinduced insulator-to-metal transition, transport measurements and x-ray photoemission spectroscopy measurements indicate that the relative stability of the metallic and insulating phases interchange around 80 K in the middle of a very wide hysteresis loop, which is a manifestation of the large potential barrier due to the long-range elastic energy. It is shown that photons are much more effective in overcoming the barrier via the electronically excited intermediate states than via the heat mode (Fig. 4.10) [14].

$Pr_{0.5}Sr_{0.5}MnO_3$ is another typical example of phase competition among different electronic states. Under cooling, two manganite transitions are observed at T_C ≈270 K from a PMI to FMM and at T_N ≈140 K to COOI in bulk ma-

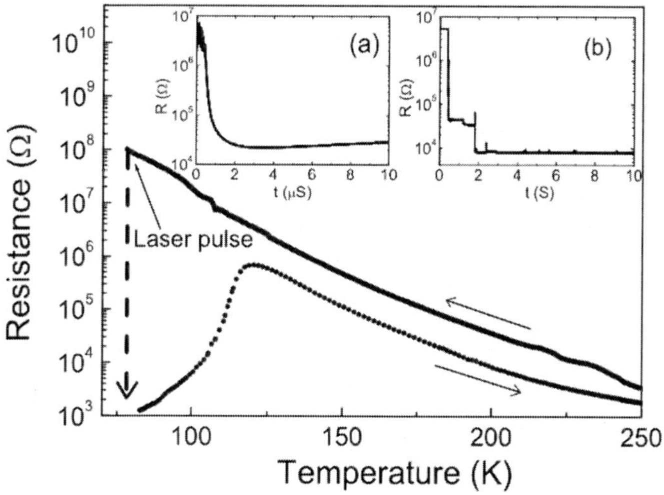

Figure 4.9. Temperature dependence of resistance for PCSMO (y=0.25) thin film. Solid and dotted lines denote the cooling and warming processes, respectively. At 77 K in the cooling process, the sample was irradiated with 100 pulses. The photon energy was 1.95 eV (λ=637 nm) and the pulse energy density was I=3.82 mJ/cm^2(Reprinted with permission from Reference [13], Copyright (2005) by the American Physical Society).

terial [15]. Hysteresis between the cooling and warming processes ensures that the FMM to COOI transition is of the first order. It is evident that the FMM and COOI states compete with each other in the thermal hysteresis region. Light irradiation and external magnetic field can be employed to disturb the competition between the FMM and COOI states and to induce metastable states. Hu *et al.* have observed thermal history dependent photoconductivity in the thermal hysteresis region of $Pr_{0.5}Sr_{0.5}MnO_3$ thin film. In the cooling process, the film shows an upward resistance relaxation in darkness. Under light illumination, the transient photoconductivity effect is observed. However, it exhibits a downward resistance relaxation without illumination and PPC behavior with light illumination in the warming process (Fig. 4.11). These phenomena can be explained by the competition and thermal history dependence of stability among the FMM and COOI states. These features may be utilized in photo-/magnetic field sensitive and memory devices [16]. For $La_{0.625-y}Pr_yCa_{0.375}MnO_3$, the

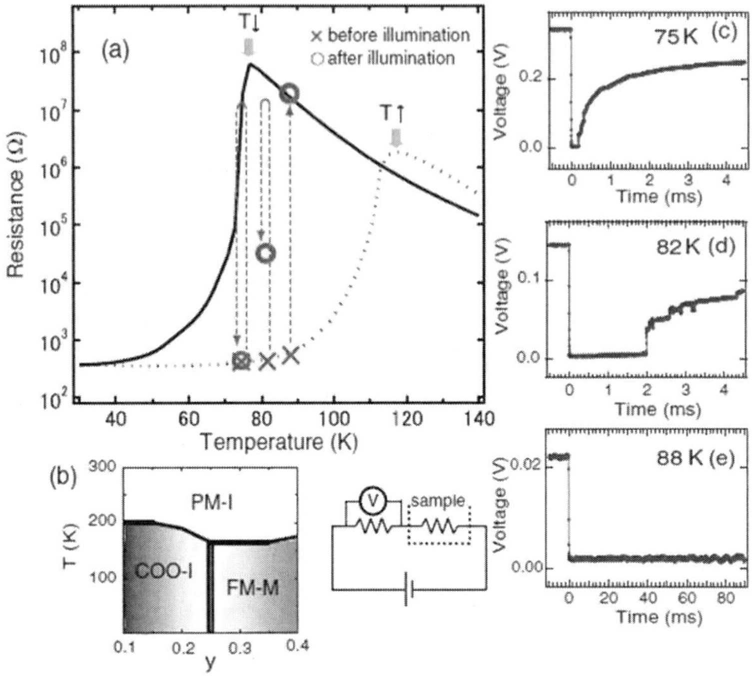

Figure 4.10. (a) Temperature dependence of the resistance for a PCSMO thin film. Solid and dotted lines denote the cooling and warming processes, respectively. The dotted arrows denote the changes caused by the laser illumination. (b) Phase diagram of PCSMO. (c)-(e) Voltage change across the reference resistance caused by the laser illumination at t=0 at (c) 75, (d) 82, and (e) 88 K(Reprinted with permission from Reference [14], Copyright (2008) by the American Physical Society).

FMM and COOI states coexist and compete with each other in the thermal hysteresis region to form a phase-separated system. The light illumination can be employed to disturb the phase competition between the FMM and COOI states and to generate non-thermally accessible metastable states. Chaudhuri *et al.* have observed a giant and persistent drop in the resistance of 5 orders of magnitude in $La_{0.175}Pr_{0.45}Ca_{0.375}MnO_3$ thin films of the insulating ground state upon photo-illumination with a continuous wave laser, a pulsed laser and a white light source. This photoinduced metallic state is eternal at zero-bias as long

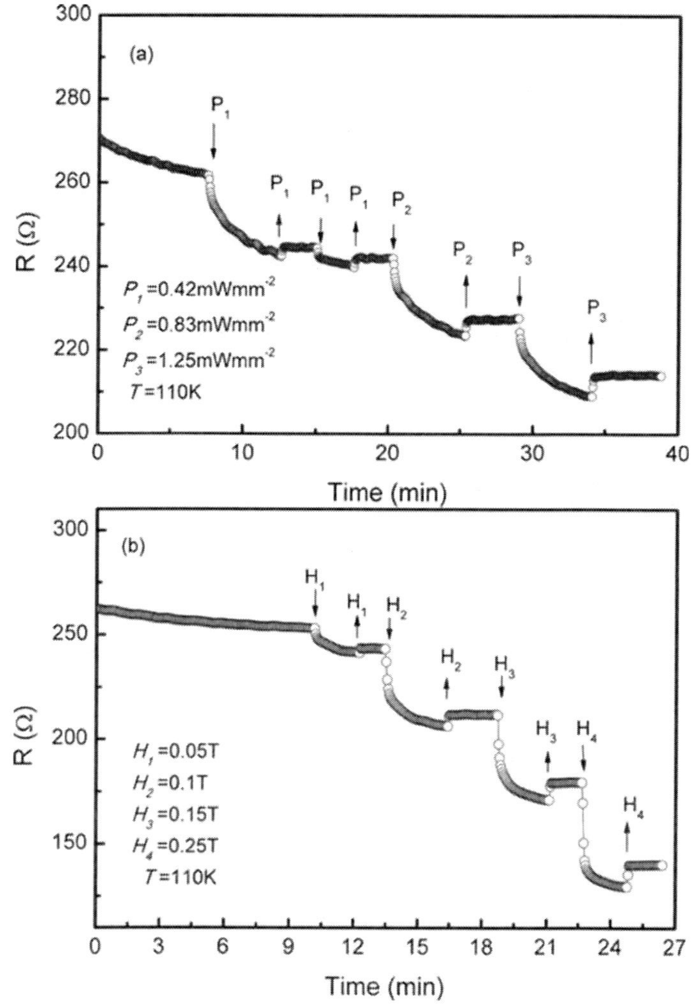

Figure 4.11. Time dependence of the resistance with (a) various light intensities and (b) various magnetic fields at 110 K in the warming process(Reproduced with permission from Reference [16]. Copyright 2009, AIP Publishing LLC).

as the sample temperature is held constant. The net drop in the resistance is a function of the photon flux and sample temperature, and can be accentuated on application of an electric field (Fig. 4.12) [17]. Hu *et al.* have observed a crossover from positive to negative PPC in the thermal hysteresis region of

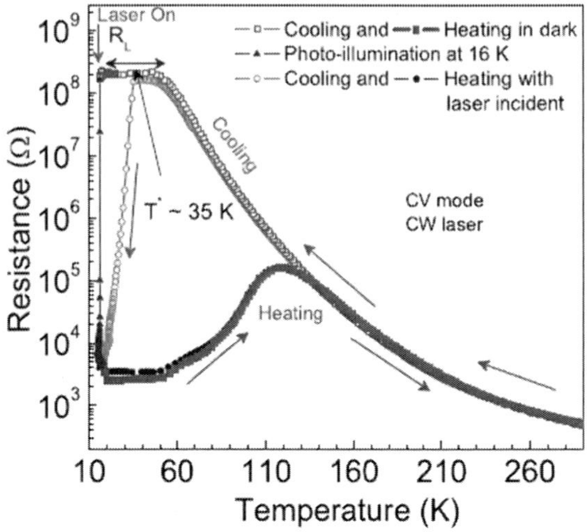

Figure 4.12. Temperature dependence of resistance of the $La_{0.175}Pr_{0.45}Ca_{0.375}MnO_3$ sample measured in a continuous wave mode for cooling in the dark (open squares) followed by photo-illumination with a laser ($h\nu$=2.8 eV and $I = 800\ mWcm^{-2}$) at 16 K (filled triangles) and finally heating back to 300 K in the dark (filled squares)(©IOP Publishing. Reproduced with permission from Reference [17]. All rights reserved).

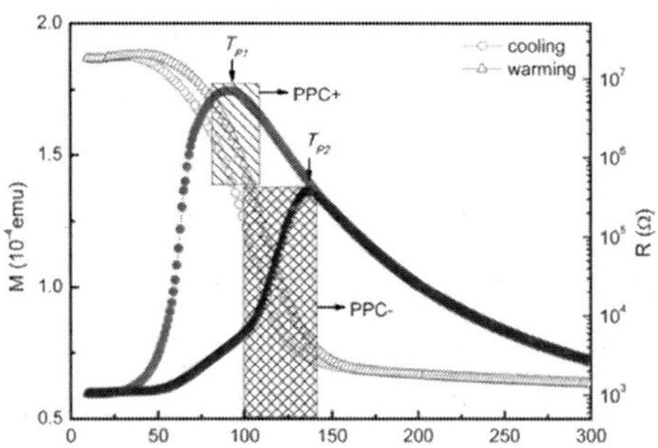

Figure 4.13. Temperature dependence of resistance and magnetization of the LPCMO thin film in both cooling and warming runs. The hatched areas represent the 'PPC+' ('PPC-') region in the cooling (warming) process(©IOP Publishing. Reproduced with permission from Reference [18]. All rights reserved).

$La_{0.325}Pr_{0.3}Ca_{0.375}MnO_3$ thin film (Fig. 4.13). In the cooling process, the resistance shows downward relaxation in darkness. Light illumination induces a resistance drop and positive PPC is observed. However, upon warming, the resistance exhibits upward relaxation without illumination. Moreover, the response of the film to light illumination shows distinct behavior. The resistance decreases to a lower value when the illumination is switched on and then shows upward relaxation during illumination. It recovers to a higher value when the illumination is stopped, which characterizes as negative PPC. These results are discussed based on the phase competition and phase stability switch between FMM and COOI states [18].

Theoretical studies of systems with competing degrees of freedom that are in a delicate equilibrium predict the formation of transient hidden phases that can be accessed by optical stimulation in the dynamical processes of photoinduced phase transitions (PIPTs) [19]. Thus, strongly correlated materials are promising in searching for PIPTs because they are likely to have hidden phases that are energetically almost degenerate but thermally inaccessible. Ichikawa *et al.* have reported the photoinduced change in the lattice structure of a charge and orbitally ordered $Nd_{0.5}Sr_{0.5}MnO_3$ thin film using picosecond time-resolved x-ray diffraction. The photoinduced state is structurally ordered, homogeneous, metastable and has crystallographic parameters different from any thermodynamically accessible state. A femtosecond time-resolved spectroscopic study shows the formation of an electronic gap in this state. In addition, the threshold-like behaviour and high efficiency in photo-generation yield of this gapped state highlight the important role of cooperative interactions in the formation process. These combined observations point towards a 'hidden insulating phase' distinct from that found in the hitherto known phase diagram (Fig. 4.14) [20].

Bulk $Bi_{0.4}Ca_{0.6}MnO_3$ has a charge-ordering temperature of 330 K. At the charge-ordering transition, the temperature dependence of the resistivity has a kink (a change in derivative) accompanied by a peak in the temperature dependence of magnetization. Smolyaninova *et al.* have reported charge-ordered $Bi_{0.4}Ca_{0.6}MnO_3$ thin films with charge-ordering temperature near room temperature, and observation of large photoinduced resistivity changes in these films associated with melting of charge ordering by visible light.The substrate-induced strain makes the charge order less stable and more susceptible to breakdown by illumination with visible light. Thus, thinner films grown on different substrates exhibit a larger drop in resistivity under illumination than thicker films. The largest photoinduced resistivity changes were observed for thin films

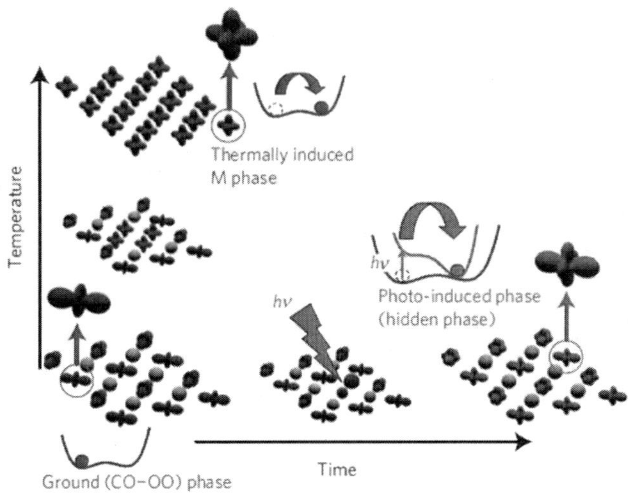

Figure 4.14. Schematic illustration contrasting the orbital character of a photoinduced hidden phase with that of a thermally induced M phase converted from the initial COOI state (ground state)(Reprinted by permission from Macmillan Publishers Ltd: Nature Materials (Ref.[20]), copyright (2011)).

of 40 nm grown under small compressive strain (on $LaAlO_3$) (Fig. 4.15)[21].

The oxygen vacancies in the doped manganites can strongly affect their physical properties. For instance, the increased oxygen deficiency in $La_{2/3}Sr_{1/3}MnO_{3-\delta}$ leads to a negative electric charge deficiency which is compensated by a Mn^{4+} decrease to keep charge neutrality. So increased oxygen vacanies also decreases the Mn^{4+}/Mn^{3+} ratio. Thus the mixed Mn valence can be changed by oxygen deficiency. The insulator-metal transition temperature (T_P) decreases with oxygen deficiency and for high oxygen deficiency, this manganite is semiconducting in the whole temperature range. In particular a related class of materials, the high-temperature superconducting cuprates exhibiting in their oxygen deficient state, PPC due to the photoinduced increase in the carrier density [22]. Perhaps more interestingly, they also exhibit persistent photosuperconductivity, a light induced collective behavior [23]. The interesting question arises whether it is possible to induce PPC and a collective magnetic effect in the manganites. Cauro *et al.* have observed that the light illumination produceds major, interesting changes of the electrical properties of oxygen de-

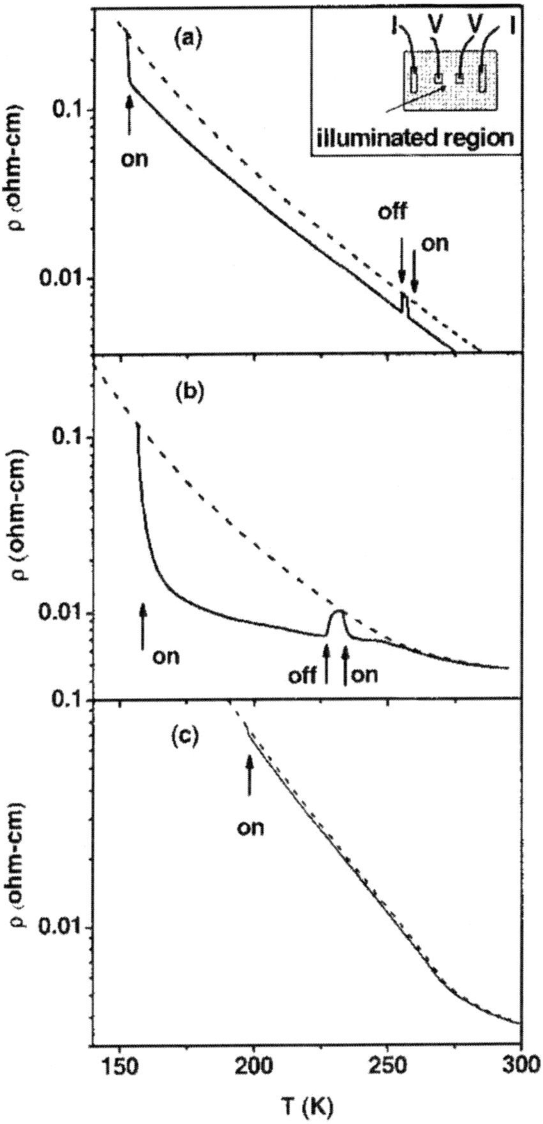

Figure 4.15. Photoinduced resistivity changes for $Bi_{0.4}Ca_{0.6}MnO_3$ thin films: (a) 40 nm on STO, (b) 40 nm on LAO, and (c) 120 nm on LAO(Reproduced with permission from Reference [21]. Copyright 2005, AIP Publishing LLC).

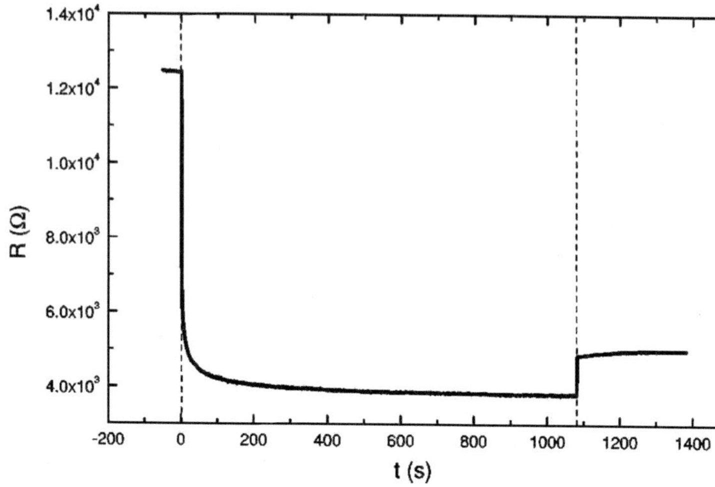

Figure 4.16. Resistance versus illumination time for the sample $La_{2/3}Sr_{1/3}MnO_{3-\delta}$ ($\delta = 0.122$) showing persistent photoconductivity (T=10 K)(Reprinted with permission from Reference [24], Copyright (2001) by the American Physical Society).

ficient $La_{2/3}Sr_{1/3}MnO_{3-\delta}$. At small oxygen deficiency, the T_P decreases with increasing oxygen deficiency. The low-temperature metallic behavior in darkness shows a persistent increase of the conductivity with light (Fig. 4.16). This photoconductivity increases with oxygen deficiency (from a few to 60%). With large oxygen deficiency, the films are semiconducting in darkness from helium to room temperature, and light induces a transient photoconductivity and a low-temperature insulator-metal transition. The PPC effect arises from an increase in the charge carrier density by light and the consequent trapping of the electrons in oxygen vacancies [24]. Sheng *et al.* have observed both transient and persistent photoconductivity in oxygen-deficient $La_{2/3}Sr_{1/3}MnO_{3-\delta}$ thin films at low temperature. To test whether the PPC effect can be influenced by applied magnetic fields, a magnetic field parallel to the film plane is applied [25].

Yuan *et al.* have reported the photoinduced switching of resistance by several orders of magnitude in oxygen-deficient $La_{0.6}Ca_{0.4}MnO_{3-\delta}$ thin films deposited on Si substrates (Fig. 4.17). The magnitude of the transient photoconductivity could attain 95.9% at room temperature under 532 nm cw laser illumination at laser intensity of 142 mW/cm^2. The switching has a fast pho-

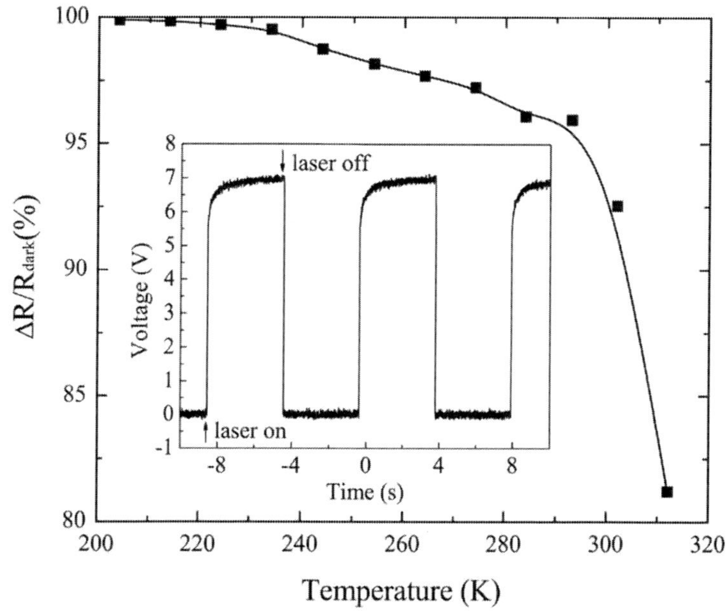

Figure 4.17. Temperature dependence of TPC magnitude under the cw laser illumination. The inset shows a repeated change of voltage drop across a reference resistor (5 kρ) with an applied voltage of 20 V by switching on and off the cw laser at room temperature(Reproduced with permission from Reference [26]. Copyright 2007, AIP Publishing LLC).

toresponse effect on a nanosecond time scale when irradiated by a 532 nm laser pulse of 7 ns duration. The photocurrent and photoresponse times depend on the laser intensity and the applied electric field [26]. Beyreuther *et al.* have studied the transport characteristics of reduced $La_{0.7}Ce_{0.3}MnO_{3-\delta}$ thin films under photoexciation and demonstrated large photoconductivity of up to seven orders of magnitude at 100 K and the recovery of the insulator-metal transition under illumination (Fig. 4.18). The strength of the effect is comparable to the CMR effect. The investigation of the temporal behavior of the photoconductivity reveals that only a small part of the effect is persistent. The spectral dependence of the photoconductivity shows several kinks attributed to the $SrTiO_3$ substrate. Thus, the substrate needs to be taken into account for the understanding of the observed photoconduction. The experimental results rule out a strong contribu-

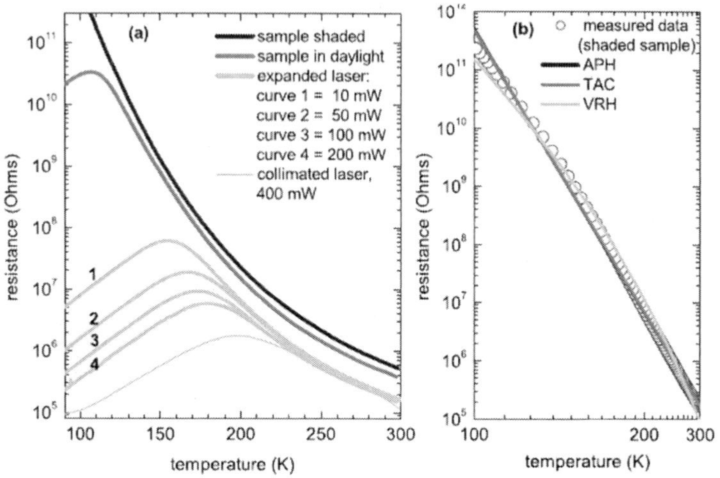

Figure 4.18. Temperature dependence of the resistance of the reduced LCeMO film in darkness (sample shaded), in diffuse daylight, and under laser illumination (λ=514 nm) at different intensities. (b) Fit of the resistance-temperature characteristics of the shaded sample by three different transport models: APH, simple TAC, and VRH(Reprinted with permission from Reference [27], Copyright (2009) by the American Physical Society).

tion from substrate photoconduction for the wavelength above 500 nm, whereas excitation across the $SrTiO_3$ band gap leads to a strong drop of the resistance attributed to carrier injection into the film and/or substrate photoconduction [27]. Smolyaninova *et al.* have found that resistivity of the $Bi_{0.4}Ca_{0.6}MnO_3$ thin films increases with oxygen depletion during being annealed in argon or air. Higher annealing temperatures lead to larger oxygen depletion and higher resistivity values. Oxygen-deficient $Bi_{0.4}Ca_{0.6}MnO_3$ films exhibit a significant increase in the photoinduced resistivity changes and the lifetime of photoinduced conductive phase (Fig. 4.19). Two mechanisms can possibly contribute to the enhancement of photoinduced effects in CO $Bi_{0.4}Ca_{0.6}MnO_3$ thin films: (1) trapping of electrons by the oxygen vacancy, while the holes (from electron-hole pair created by incoming photon) contribute to conduction; and (2) weakening of the charge ordering by the disorder introduced by the oxygen vacancies, so it can be more easily destroyed by light [28].

Hu *et al.* have investigated the effects of light and electrical current on

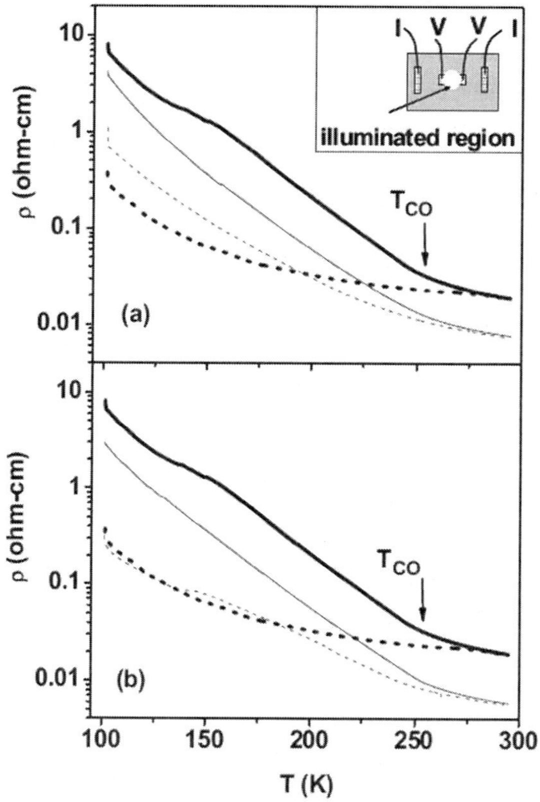

Figure 4.19. Temperature dependence of the resistivity of the $Bi_{0.4}Ca_{0.6}MnO_3$ thin film grown on LAO: solid lines-taken on cooling without illumination; dashed lines-taken on warming with illumination for the (a) as-grown film (thin lines) and film annealed in argon (thick lines); for the (b) film annealed in argon (thick lines) and film annealed in oxygen after annealing in argon (thin lines)(Reproduced with permission from Reference [28]. Copyright 2009, AIP Publishing LLC).

the electrical transport properties of oxygen-deficient $La_{7/8}Sr_{1/8}MnO_{3-\delta}$. It is found that the photoconductivity can be remarkably enhanced by increasing the electrical current. Figure 4.20 shows the temperature dependent resistance of the oxygen-deficient LSMO thin film under light illumination and various electrical currents. In darkness, the film is highly resistive and exhibits insulating

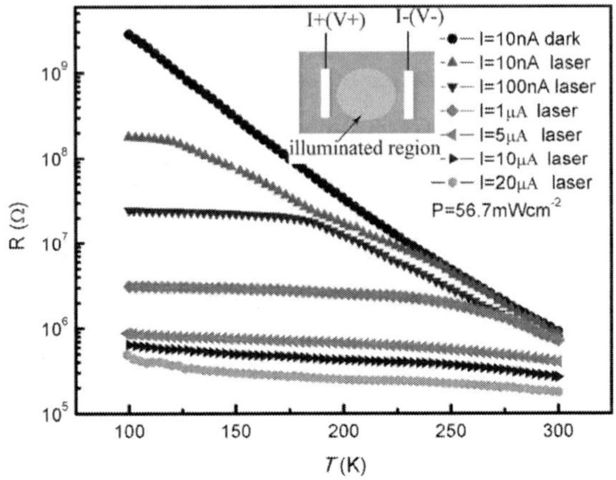

Figure 4.20. Temperature dependence of resistance at various electrical currents and light illuminations of the oxygen-deficient LSMO thin film. Inset shows the experimental current and voltage contact configuration with respect to the illuminated region(Reproduced with permission from Reference [29]. Copyright 2012, AIP Publishing LLC).

behavior. Under light illumination of 56.7 $mWcm^{-2}$, the resistance drop of the film is more than one order of magnitude at 100 K when the electrical current is 10 nA. It is interesting to note that the photoconductivity can be strongly enhanced by increasing the electrical current. A decrease in resistance by more than three orders of magnitude is observed at 100 K and the photoconductivity becomes evident in the whole studied temperature range when the electrical current is increased to 20 μA. The CEPC ratio achieves 80% at 300K with light intensity of 56.7 $mWcm^{-2}$ and electrical current of 20 μA. The phenomena are explained by a photoinduced localized insulator-metal transition and stabilization of conducting paths by the electrical current [29].

2.1. Photoinduced Demagnetization Effect

Zhao *et al.* have observed the pulsed laser excitation induced conductance changes in CMR material $La_{0.7}Ca_{0.3}MnO_3$ on the picosecond time scale. The femtosecond optical excitation at 1.5 eV photon energy gives rise to dramatically different photoresponses in the paramagnetic and ferromagnetic states of the manganites, resulting in a conductive and a resistive transient, respectively.

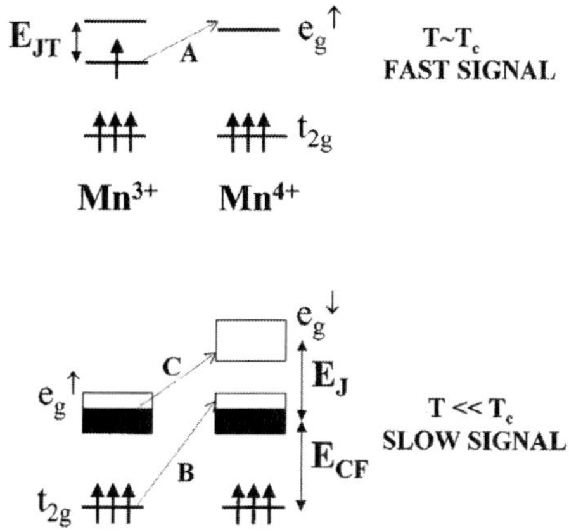

Figure 4.21. The schematic electronic structure of the e_g and t_{2g} levels of Mn^{3+} and Mn^{4+} ions and the optical transitions relevant to the experiment(Reprinted with permission from Reference [30], Copyright (1998) by the American Physical Society).

The former arises from the photoionization of a J-T small polaron, while the latter is associated with the reduction of the electron mobility in the metallic state due to long lived spin excitations (Fig. 4.21). The conductive transient measures the thermalization and localization time of an electron made itinerant by a photon, and this is found to be 150 ps and nearly independent of temperature. The resistive transient at the lowest temperatures, where the J-T effects are not manifested, decays slowly with nanosecond response and this decay time is further enhanced over an order of magnitude at higher temperatures where the J-T effects begin to dominate [30].

Liu *et al.* have investigated the effects of pulse-laser irradiation on $La_{0.7}Ca_{0.3}MnO_3$ film by the pump and probe technique as function of temperature. The observed temperature-dependent photoinduced transmission changes at E_{probe}=1.55 eV and 2.54 eV can be interpreted in term of creation ($T \leqslant T_C$) or annihilation ($T \geqslant T_C$) of the CO clusters (Fig. 4.22) [31]. Moshnyaga *et al.* have observed the increase of the resistance up to two orders of manganite under laser illumination (λ=760 nm) in $La_{0.7}Ca_{0.3}MnO_3$ epitaxial thin films in the FMM state (Fig. 4.23). Optical absorption also increases by 10-15% and the magnetic second-harmonic generation signal decreases down to zero under the irradiation. The light induced changes are reversible with characteristic

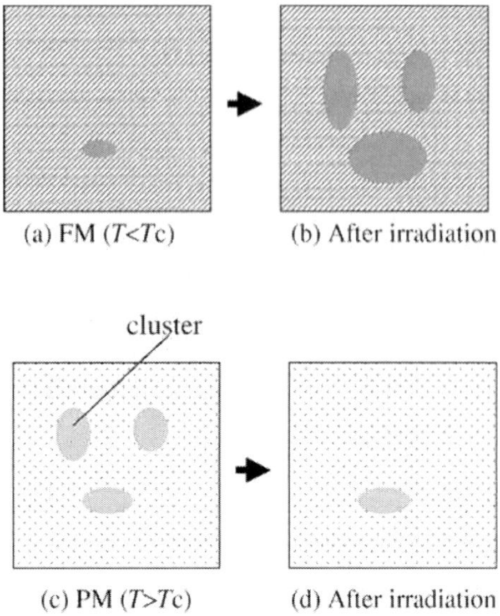

Figure 4.22. Schematic pictures of photoirradiation effects:(a) before and (b) after photoirradiation in the ferromagnetic ($T < T_C$) state and, (c) before and (d) after photoirradiation in the paramagnetic ($T > T_C$) state(Reprinted with permission from Reference [31], Copyright (2001) by the American Physical Society).

relaxation times $t \sim 1 - 30$ s. Magnetic field, B=4 T, suppresses the photoconductivity and decreases its relaxation time. Photoinduced effects are caused by the injection of a large number of extra carriers, which change the AFM/FM phase balance in $La_{0.7}Ca_{0.3}MnO_3$, favoring the insulating AFM state [32]. Hu *et al.* have investigated the effects of magnetic field and light illumination on the electrical transport properties of $La_{7/8}Ca_{1/8}MnO_3$ thin film. The film shows an insulator-metal transition at $T_P \approx 191.9K$ and a low-temperature resistance minimum at $T_{min} \approx 48K$ in darkness. Both magnetic field and light illumination shift the T_P to higher temperatures. However, the magnetic field shifts the T_{min} to a lower temperature and the light illumination shifts the T_{min} to a higher one (Fig. 4.24). These phenomena could be explained by the photoinduced hole-doping and demagnetization effects induced by the light illumination [33]. Smolyaninova *et al.* have observed a significant photoinduced increase of resistance in FMM state of $La_{0.7}Ba_{0.3}MnO_3$ thin films consistent with

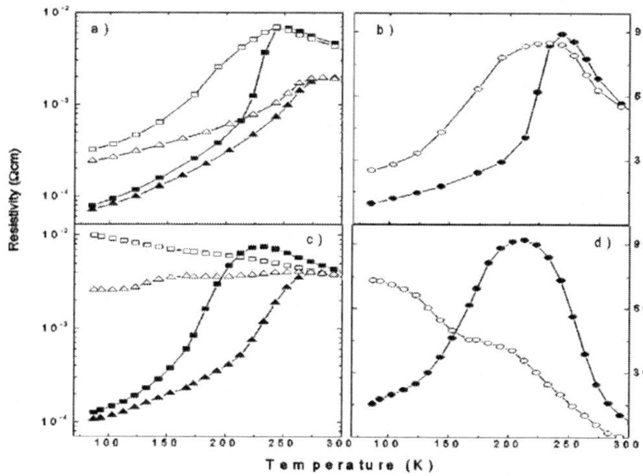

Figure 4.23. Electrical resistivty ρ and CMR as a function of temperature for a LCMO film with thickness 120 nm [(a) and (b)], and for a trilayer structure LCMO(15 nm)/MgO(7 nm)/LCMO(15 nm)[(c) and (d)]. Closed and open symbols denote ρ values in dark and under illumination, respectively. Square symbols refer to the ρ values at zero (ambient) magnetic field, B=0, and triangles denote ρ (B=4 T). Closed and open circles relate to the CMR data in dark and under irradiation, respectively(Reproduced with permission from Reference [32]. Copyright 2004, AIP Publishing LLC).

photoinduced demagnetization. These photoinduced effects strongly depend on thermal/illumination history and are associated with the energy balance between FMM and PIR phases at different temperatures (Fig. 4.25) [34]. Guo *et al.* have performed systematic studies on the effects of magnetic field on the photoinduced demagnetization effect in $La_{0.8}Ca_{0.2}MnO_3$ films and observed that the photoinduced demagnetization effect is weakened by the magnetic field. In the FMM phase, the decrease of magnetization is observed under light illumination, indicating that the photo-excitation of spin polarized electrons from the localized t_{2g} state to the travelling e_g state can decrease the magnetization of LCMO thin films. It has also been found that the PR can be decreased by external magnetic field. This behavior can be explained by the following two effects: (i) the reorientation of the FMM domains and spin direction of photoexcited carriers in LCMO films and (ii) the release and recombination of electrons trapped around

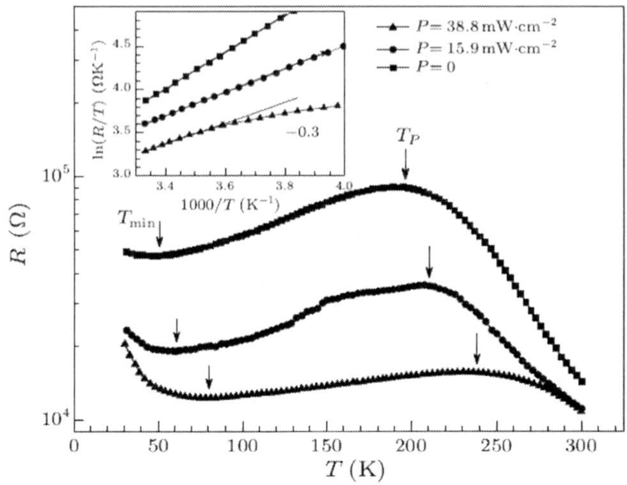

Figure 4.24. Temperature dependence of resistance of the LSMO thin film under light illumination of 0, 15.9 and 38.8 mWcm^{-2}(Reprinted with permission from Reference [33], Copyright (2010) by the Chinese Physical Society).

oxygen vacancies under magnetic field [35].

2.2. Optical-magnetic Effect

Light illumination can not only strongly influence the electrical transport properties of the doped manganites, but also affect their magnetic properties. Matsuda *et al.* have observed a significant change of absorption spectrum in the visible region by the irradiation of an intense laser pulse in a double-exchange ferromagnet $(Nd_{0.5}Sm_{0.5})_{1-x}Sr_xMnO_3$ (x=0.4) thin film. In the ferromagnetic state, the increase of spectral weight of the interband transition between J-split bands has been observed, indicating that the photoexcited e_g carriers suppress the neighboring spin correlation (Fig. 4.27). The time-resolved spectroscopy in a vast time domain from 100 fs to 1 μs has shown that the photoinduced demagnetization occurs within 200 ps after the photoinjection and spin correlation recover in 300 ns [36]. Baran *et al.* have studied the effect of light of He-Ne (λ=633 nm) and Ar (λ=488 nm) lasers on the magnetization of epitaxial films of $Pr_{0.6}La_{0.1}Ca_{0.3}MnO_3$. It has been found that the illumination stimulates the antiferromagnetic-ferromagnetic phase transition observed in this mangan-

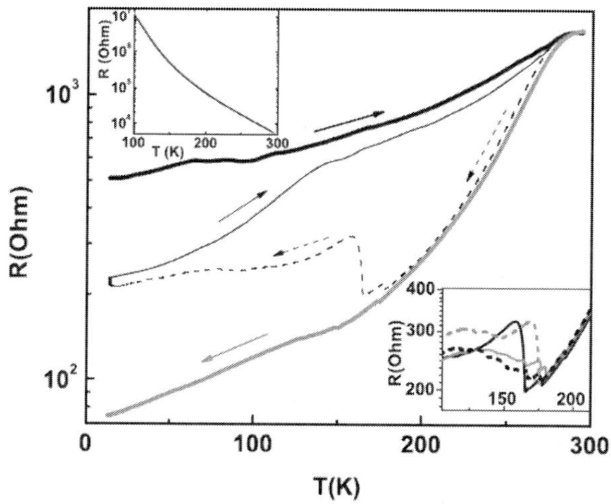

Figure 4.25. Temperature dependence of the resistance of the $La_{0.7}Ba_{0.3}MnO_3$ thin film on NGO: thick solid gray line-cooling without illumination; thick solid black lin-warming with illumination after cooling without illumination; thin dashed line-cooling with illumination; thin solid line-warming with illumination after cooling with illumination(Reproduced with permission from Reference [34]. Copyright 2011, AIP Publishing LLC).

ite in a magnetic field at low temperature. In a magnetic field, the light induces the magnetization increase that apparently is related to a partial transformation of an antiferromagnetic phase into a ferromagnetic one and a formation of ferromagnetic clusters within the antiferromagnetic phase. It was observed that a linearly polarized light affects more strongly on the magnetic state of $Pr_{0.6}La_{0.1}Ca_{0.3}MnO_3$ films than an unpolarized light (Fig. 4.28) [37]. Huhtinen *et al.* have investigated the microwave losses due to the conductivity and magnetic permeability in $La_{0.9}Ca_{0.1}MnO_3$ films illuminated with the energy range of $E = 0.5 - 2eV$. Growth of photoinduced ferromagnetic hysteresis is observed when the magnetic field is swept between -15 and 15 mT at temperatures below 60 K (Fig. 4.29). The time required for saturation of the opening of the hysteresis loop depends on the photon energy having the minimum of 40 s at the illumination intensity $I = 3.5 \times 10^{14}$ photons/cm^2. Both photoinduced magnetization and the increase of microwave photoconductivity can be well explained with a model assuming that small ferromagnetic regions exist

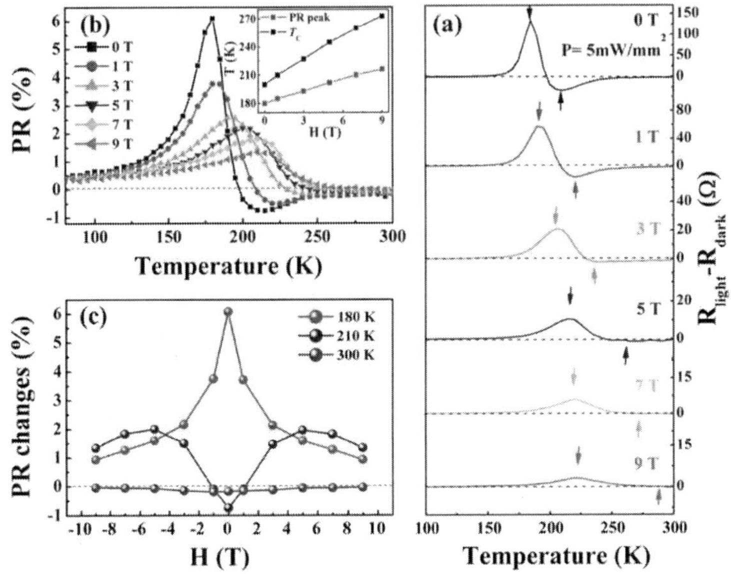

Figure 4.26. The temperature dependence of (a) $R_{light} - R_{dark}$ and (b) PR for LCMO film under various fields of 0, 1, 3, 5, 7, and 9 T, respectively. The arrows in (a) point out the positions of the maximum and minimum PR. The inset of (b) shows the field dependence of the metal-to-insulator transition temperatures TC and the temperatures at positive PR peaks. (c) The field dependence of PR measured in 180, 210, and 300 K, respectively(Reproduced with permission from Reference [35]. Copyright 2012, AIP Publishing LLC).

within an insulating ferromagnetic phase of the sample and that these regions are expanded by optically induced charge transfer between Jahn-Teller split e_g states of neighboring Mn^{3+} ions [38].Okimoto *et al.* have showed that persistent ferromagnetic magnetization as large as 0.6 μ_B per Mn site can be induced by irradiating visible laser pulses on a thin film of 1% Cr-doped $Pr_{0.5}Ca_{0.5}MnO_3$. The enhancement of M upon the photoexcitation is also confirmed by the scanning superconducting quantum interference device microscope as an increase of magnetic field from ferromagnetic domains. A color bar denotes the magnitude of out-of-plane magnetic filed (B_z). Before the irradiation, the value of B_z is small with rather weak contrast of blue and red, which is partly due to the coexistence of the CO and FM domains and magnetic domains within FM

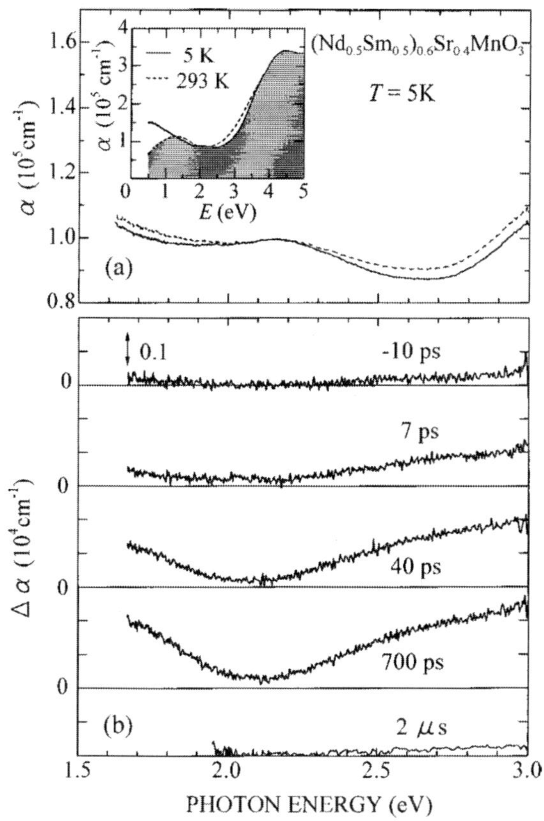

Figure 4.27. (a) Absorption spectra for the $(Nd_{0.5}Sm_{0.5})_{0.6}Sr_{0.4}MnO_3$ thin film before (solid curve) and after pump-pulse excitations (dashed curve).(b) Differential absorption spectra $\Delta\alpha(\omega)$ in ferromagnetic phase at 5 K. The excitation photon energy and density are 3.20 eV and $\sim 8\times10^{19}$ cm^{-3}, respectively. Delay times between pump and probe pulses are shown in the figure(Reprinted with permission from Reference [36], Copyright (1998) by the American Physical Society).

regions. After the illumination, the contrast of the image becomes clearer, and the amplitude of B_z of the magnetic domains is remarkably enhanced up to $\approx$10 mT [39].

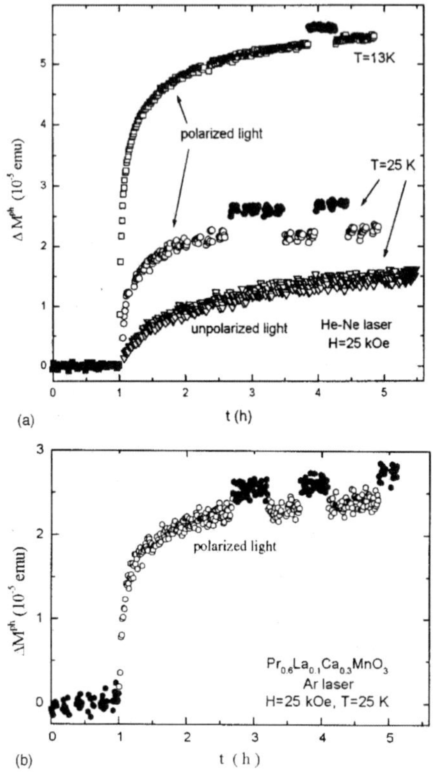

Figure 4.28. The time evolution of magnetization at 25 kOe (illumination was turned on at t=1 h). Open and full dots represent data obtained under and without illumination, respectively. (a) Illumination by He-Ne laser: linearly polarized light (circles: 25 K, squares: 13 K), unpolarized light (triangles: 25 K). (b) Illumination at 25 K with linearly polarized light of Ar laser(Reprinted with permission from Reference [37], Copyright (1999) by the American Physical Society).

3. Photoinduced Effects in Manganite-based Heterojunctions

3.1. Photovoltaic Effects in Heterojunctions

In conventional semiconductor technology, it is convenient to modulate the carrier concentration of heterostructures through the interface design of band struc-

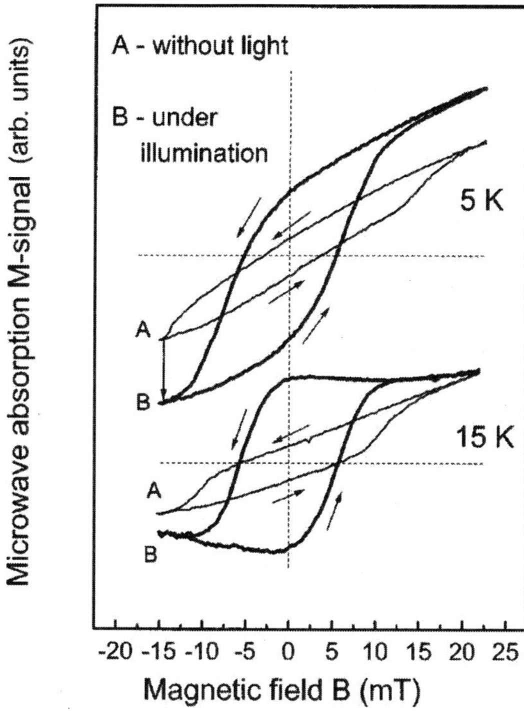

Figure 4.29. Hysteresis of the microwave absorption M signal without illumination (A) and under illumination (B), measured at T=5 and 15 K(Reprinted with permission from Reference [38], Copyright (2000) by the American Physical Society).

ture. In the *p-n* junction, for example, a depletion layer and built-in electric field are formed at the interface so as to unify the Fermi level all over the junction. Analogously, transition-metal-oxide compounds, as emergent electronics materials with correlated electrons, often exhibit nontrivial and potentially useful electronic properties at heterointerfaces. Moreover, the charge state in the strongly correlated electron oxides is closely coupled with the spin and orbital degrees of freedom. Therefore, the doped manganites provide a challenging arena to explore new multifunctional materials. With this notion in mind, it is of interest to investigate the properties of the manganite-based heterojunction under electric field, magnetic field and light aiming at the exploration of novel

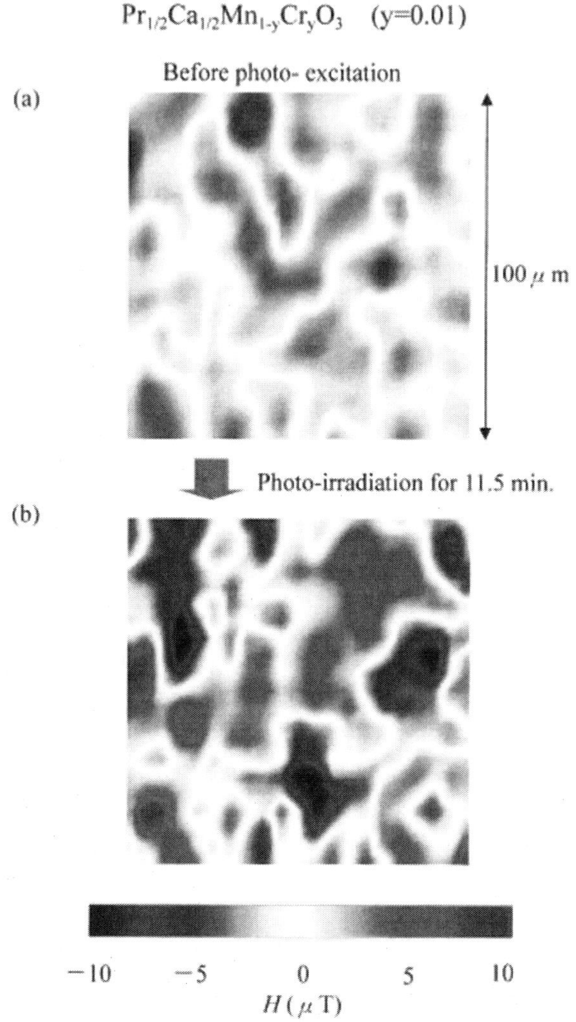

Figure 4.30. Images of magnetic domain structures at 3 K in a $Pr_{0.5}Ca_{0.5}Mn_{0.99}Cr_{0.01}O_3$ thin film taken by scanning SQUID microscope (a) before and (b) after the irradiation of frequency-doubled cw NdCYVO4 laser (532 nm) for ≈11.5 min with a power density of ≈0.5 mW/mm^2. A color bar denotes the magnitude of out-of-plane magnetic field(Reproduced with permission from Reference [39]. Copyright 2002, AIP Publishing LLC).

physical phenomena and theory.

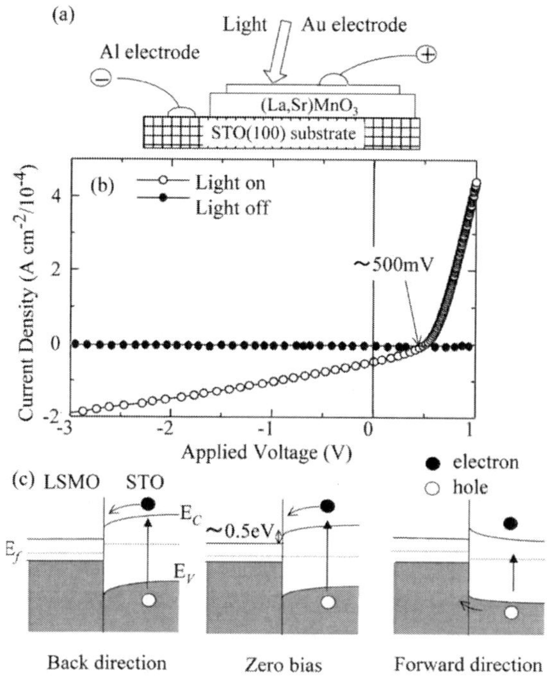

Figure 4.31. I-V curve of the LSMO/STO heterostructure in the perpendicular direction measured at 30 K. The upper image shows a schematic illustration of the I CV measurement. (c) Band diagram at the interface of the LSMO/STO heterostructure(Reproduced with permission from Reference [40]. Copyright 2000, AIP Publishing LLC).

Katsu *et al.* have fabricated $La_{1-x}Sr_xMnO_3/SrTiO_3$ heterojunction and observed the photovoltage of 500 mV by irradiation of light at a wavelength below 400 nm. Due to the high resistivity of $SrTiO_3$, an electric current has not been detected without light irradiation. With irradiation, the pohtocarrier is generated in $SrTiO_3$ and the photovoltage of 500 mV is detected (Fig. 4.31) [40]. Sun *et al.* have fabricated a *p-n* junction of $La_{0.29}Pr_{0.38}Ca_{0.33}MnO_3$/STO:Nb (0.5 wt%Nb-doped $SrTiO_3$) and have found the heterojunction exhibited a significant photovoltaic effect and a somewhat acceptable rectifying behavior (Fig. 4.32) [41]. In addition, the photovoltaic effect in $La_{0.7}Ce_{0.3}MnO_3$/STO:Nb het-

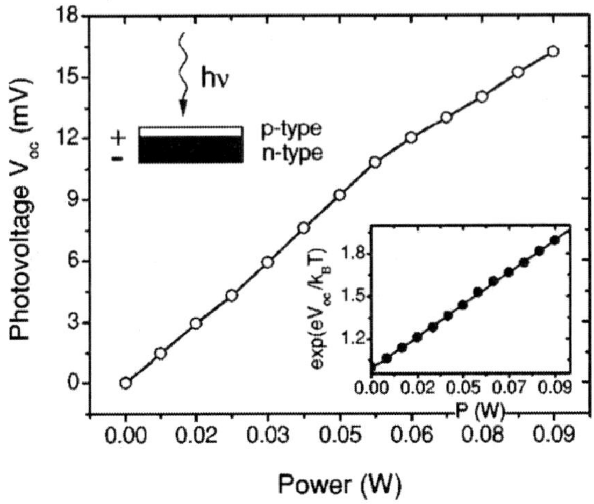

Figure 4.32. Photovoltage as a function of light power (λ=532 nm) measured at room temperature(Reproduced with permission from Reference [41]. Copyright 2004, AIP Publishing LLC).

erojunction and rectifying and photovoltaic properties in $CaMnO_3$/STO:Nb heterojunction was observed [42, 43]. Sun *et al.* also observed a strong temperature dependence of photovoltaic effect in $La_{0.32}Pr_{0.35}Ca_{0.33}MnO_3$/STO:Nb heterojunction. The synchronous variation in the photovoltage and the magnetization $La_{0.32}Pr_{0.35}Ca_{0.33}MnO_3$/STO: Nb heterojunction indicates the magnetic origin of the temperature dependence of the photovoltaic effect (Fig. 4.33). It implies that the temperature affects the photovoltaic effect by modifying the magnetic order, and then the band structure of $La_{0.32}Pr_{0.35}Ca_{0.33}MnO_3$ [44]. Sun *et al.* have observed photovoltaic effect in $La_{0.29}Pr_{0.38}Ca_{0.33}MnO_3$/STO:Nb *p-n* junction with a photovoltaic pulse of ,8 ms full width at half-maximum (FWHM) when the junction was irradiated by a 532 nm laser pulse of 10 ns duration [41]. Lu *et al.* have observed ultrafast photoelectric effects in $La_{0.7}Sr_{0.3}MnO_3$/Si *p-n* junctions. The rise time was ∼210 ps and the FWHM was 650 ps for the photovoltaic pulse when the junction was irradiated by a 1064 nm laser pulse of 25 ps duration [45].

It is known that the magnetic behavior of manganite films can be changed by external magnetic fields. Therefore, it should be of significance to check whether the observed photovoltaic effect can be manipulated by the external

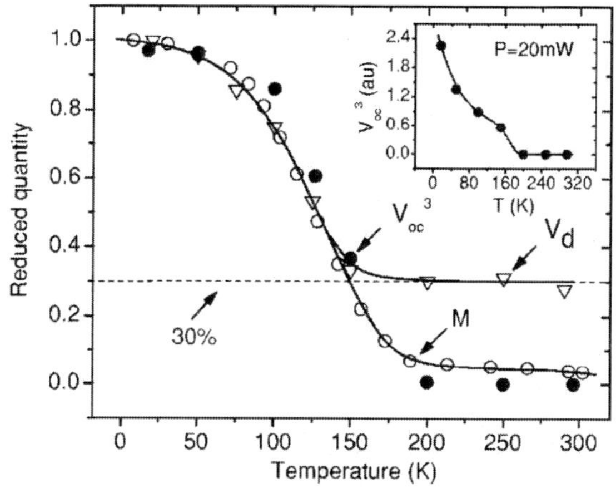

Figure 4.33. Comparison of the normalized $V_d(T)$, $M(T)$, and $V_{oc}^3(T)$ relations. The $M(T)$ curve was measured under a field of 0.1 T in the field-cooling mode, and the $V_{oc}(T)$ curve was obtained under the light intensity of 4 mW. The inset plot shows the photovoltage corresponding to the light intensity of 20 mW. Solid lines are guides for the eye, and the dashed line marks the expected threshold for magnetic percolation(Reproduced with permission from Reference [44]. Copyright 2004, AIP Publishing LLC).

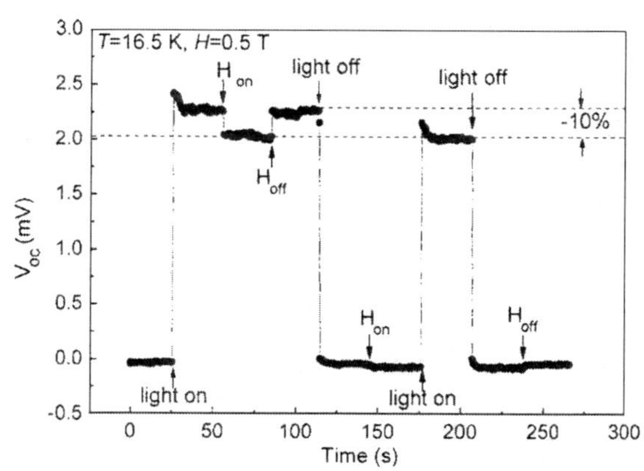

Figure 4.34. Response of photovoltage to light illumination measured at a constant temperature of 16.5 K with the light power density of 0.34 mW/mm^2 and at an applied magnetic field of 0.5 T parallel to the film surface(Reproduced with permission from Reference [46]. Copyright 2005, AIP Publishing LLC).

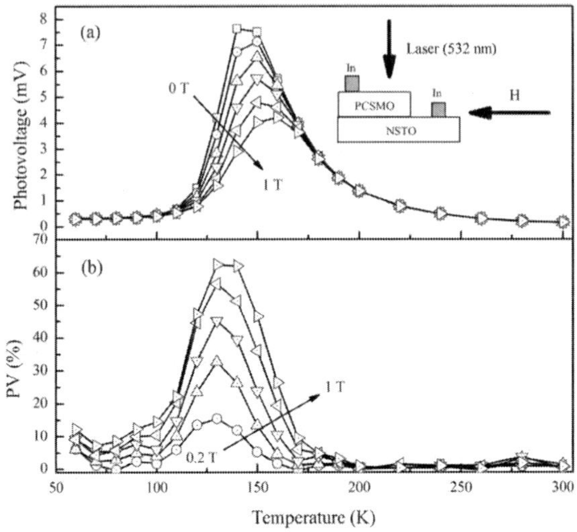

Figure 4.35. Temperature dependence of (a) photovoltage of heterostructure irradiated by a laser (15 mW/mm^2) and (b) PV for PCSMO/NSTO under different magnetic fields, respectively(Reproduced with permission from Reference [47]. Copyright 2013, AIP Publishing LLC).

magnetic field. Sheng *et al.* have investigated the effect of magnetic field on the photovoltaic effect for a manganite-based heterojunction composed of a $La_{0.7}Ce_{0.3}MnO_3$ film and a 0.5 wt % Nb-doped $SrTiO_3$ substrate. A linear decrease in the photovoltage with magnetic fields is observed and the relative variation of photovoltage is larger than 10% under a field of 0.5 T at a temperature T = 16.5 K (Fig. 4.34). The modulation of photovoltaic effect can be ascribed to the variation of magnetization and resistance of depletion layer caused by external magnetic fields [46]. Luo *et al.* have fabricated the heterostructure composed of $Pr_{0.65}(Ca_{0.75}Sr_{0.35})_{0.45}MnO_3$ and 0.7 wt % Nb-doped $SrTiO_3$. The heterostructure exhibits an asymmetric current-voltage relation similar to that of *p-n* junctions and an obvious photovoltaic effect with its maximum value of about 25.1 mV at T=140 K. It is interesting that the magnetic field has a strong suppression effect on photovoltaic effect, and the maximum relative change of photovoltage under the magnetic field of 1 T is about 63% with the illumination of 15 mW/mm^2 light (532 nm) at T=140 K (Fig. 4.35). The physical mechanism is explained by the electronic phase separation mechanism [47]. Sheng *et*

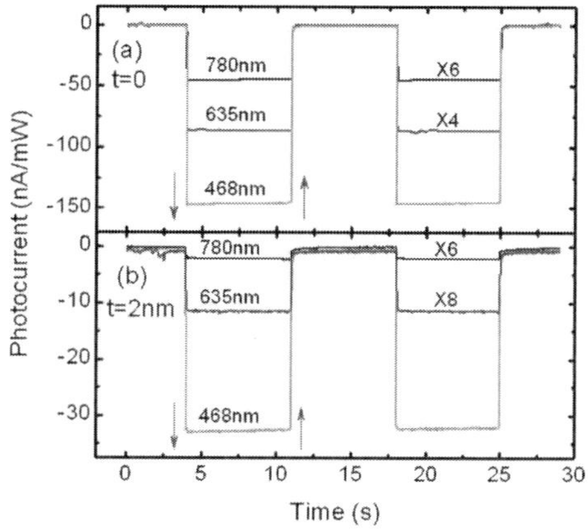

Figure 4.36. Selected photocurrent of the LCMO/LMO(t) /STON junctions with t=0 (a) and 2 nm (b), measured under the light power of 1 mW and different wavelengths(Reproduced with permission from Reference [53]. Copyright 2009, AIP Publishing LLC).

al. have explored the magneto-tunability of a pn junction based on a correlated electron oxide $La_{0.7}Sr_{0.3}MnO_3$. It is found that the photocurrent generated in the junction is enhanced by a factor of > 3 by applying a crystallographic distortion to the LSMO. Simultaneous application of an external magnetic field of 6 T can further enhance the photocurrent by 30%. The analysis reveals that the electron correlation in LSMO, tuned by an external field, has a considerable effect on the quasi-band profile of the heterointerface, and hence on the collection probability of minority carriers [48].

It is well established that the properties of the manganite junction is exclusively determined by interfacial barrier which is sensitive to interface state, both structural and electronic states [49, 50, 51, 52]. Therefore, it is expected that the changes of interface state of the manganite junction can strongly influence the photovoltaic effect. Wei *et al.* have studied the influence of $LaMnO_3$ layer, 0-12 nm in thickness, on the photovoltaic effect for the $La_{0.67}Ca_{0.33}MnO_3/LaMnO_3/SrTiO_3$:Nb junction. Presence of $LaMnO_3$ causes an obvious weakening of the photovoltaic effect, demonstrated by the reduction in the carrier density excited by each photon (Fig. 4.36). The interfacial barrier deduced from the photovoltaic effect shows a rapid growth, from ~1.22 to ~1.45 eV, as the layer thickness increases from 0 to 2 nm, and saturates at ~1.5

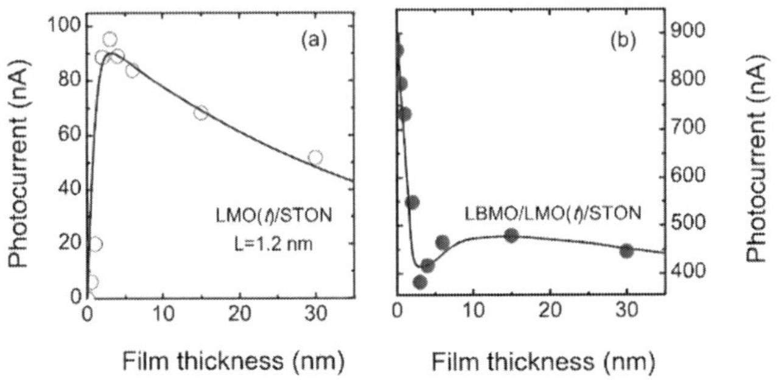

Figure 4.37. Photocurrent as a function of LMO layer thickness for the LBMO/ LMO(t)/STON (a) and LMO(t)/STON (b) junctions(Reproduced with permission from Reference [54]. Copyright 2013, AIP Publishing LLC).

eV above 2 nm. These results can be quantitatively understood assuming the occurrence of a notch and a spike in the conduction band edges at the interface of the junction [53]. Gao *et al.* have studied the photoelectronic properites of the $La_{0.67}Ba_{0.33}MnO_3$(20 nm)/$LaMnO_3$(t)/$SrTiO_3$:Nb junction with $0 \leq t \leq 30$ nm. The short-circuit photocurrent is found to show a complex dependence on the LMO buffer layer. It undergoes firstly a sharp drop as the layer thickness of LMO increases from 0 to 3 nm and then, after a rigid turn, a slow decrease for further increase in layer thickness (Fig. 4.37). The photocurrent is further found to be temperature dependent, increasing monotonically upon cooling, and the maximal growth, occurring in the junction of t=3 nm, can be as high as 226% when cooled from 320 K to 40 K. Analysis of the capacitance-voltage relations indicates that the change of interfacial barrier is the reason for the peculiar photoelectronic behavior observed [54].

3.2. Photoinduced Effects on the Physical Properties of Manganite-based Heterojunctions

Katsu *et al.* have found that the increased resistivity and a decrease of the metal-to-insulator transition temperature by 10 K occur in $La_{1-x}Sr_xMnO_3/SrTiO_3$ heterojunction under illumination of light at a wavelength below 400 nm (Fig. 4.38). An additional photoinduced metal-semiconductor transition is observed below 100 K with decreased magnetization. This behavior can be explained by considering that the hole concentration in the $La_{1-x}Sr_xMnO_3$ layer decreases

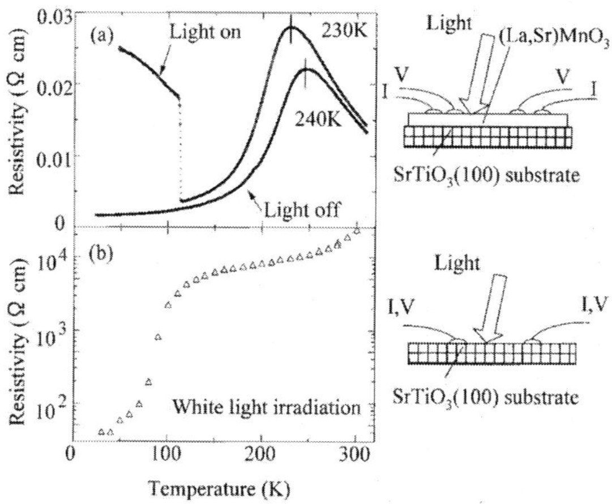

Figure 4.38. (a) Temperature dependence of resistivity of the LSMO/STO heterostructure. (b) Temperature dependence of resistivity of the STO single crystal under light irradiation. A schematic illustration of the LSMO/STO heterostructure and the STO substrate for the measurement are shown(Reproduced with permission from Reference [40]. Copyright 2000, AIP Publishing LLC).

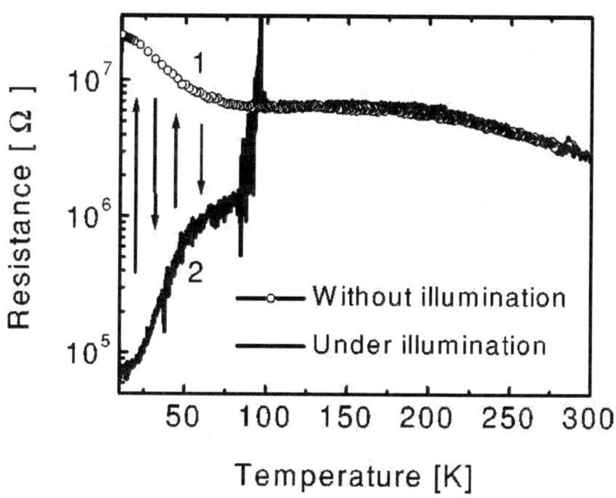

Figure 4.39. $R - T$ characteristics of a $La_{0.81}MnO_3/Al_2O_3/Nb$ tunnel junction without (curve 1) and with (curve 2) illumination(Reproduced with permission from Reference [55]. Copyright 2001, AIP Publishing LLC).

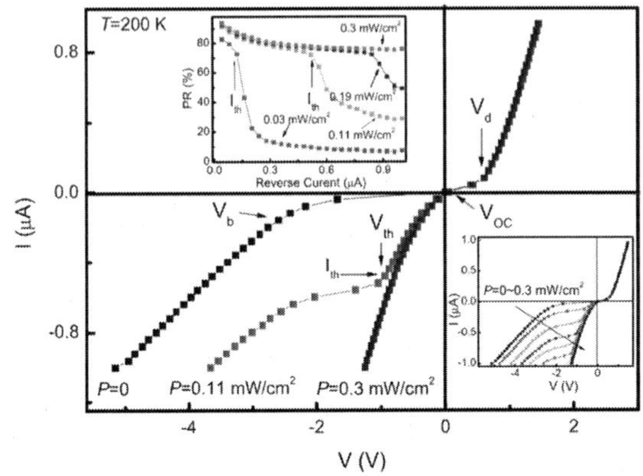

Figure 4.40. $I - V$ characteristics with and without light illumination of the PCMO/SiO_2/Si SIS junction measured at T =200 K(©IOP Publishing. Reproduced with permission from Reference [56]. All rights reserved).

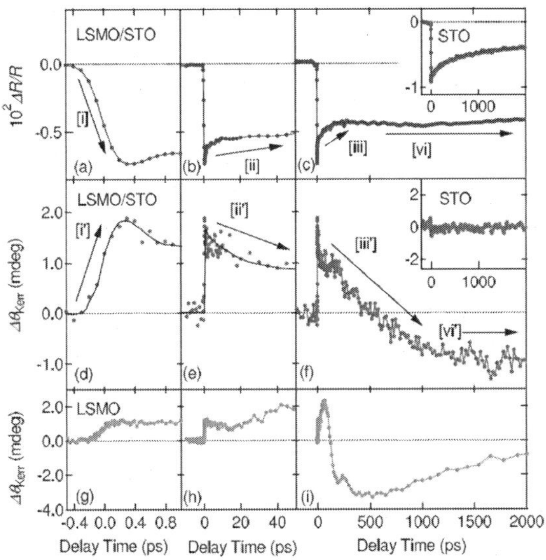

Figure 4.41. Time evolutions of (a)-(c) $\Delta R/R$ and (d)-(f)$\Delta\theta_{Kerr}$ at 80 K of a $La_{0.9}Sr_{0.1}MnO_3$/ $SrTiO_3$ heterostructure and those of (g)-(i) $\Delta\theta_{Kerr}$ at 80 K of a $La_{0.9}Sr_{0.1}MnO_3$ film(Reprinted with permission from Reference [57], Copyright (2011) by the American Physical Society).

owing to electron injection following the photoexcitation of $SrTiO_3$ [41]. Gilabert *et al.* have studied the effect of illumination by ultraviolet and visible light on the resistance versus temperature and current versus voltage characteristics of $La_{0.81}MnO_3/Al_2O_3/Nb$ tunnel junctions. Under illumination, the resistance of tunnel junctions follows the resistance in darkness at high temperatures but at temperatures lower than 95 K, the illumination induces a colossal decrease of the resistance (Fig. 4.39). This is a transient effect because the changes are reversible when the light is switched on and off [55]. Sheng *et al.* have fabricated a $Pr_{0.7}Ca_{0.3}MnO_{3-\delta}/SiO_2/Si$ heterojunction with good rectifying characteristic. Moreover, it is found that the electrical transport under forward bias is insensitive to light illumination, while that under reverse bias is sensitive (Fig. 4.40). The photoinduced resistance ratio can even reach $\sim$93.8% with a reverse bias $I = 0.04\mu A$ and is always over 75% within the whole measured current range as the light power density $P = 0.3 mWcm^{-2}$ at $T = 200$ K. The observed phenomenon is discussed according to the effect of photocarrier injection to the junction under different direction bias voltages [56]. Yada *et al.* have investigated photocontrol of magnetization in oxide heterostructures consisting of a perovskite manganite $La_{0.9}Sr_{0.1}MnO_3$ and a titanate $SrTiO_3$ by time-resolved magneto-optical Kerr effect measurements. On photoexcitation of $SrTiO_3$, the magnetization of the manganite increases within 0.2 ps and persists for $\sim$500 ps (Fig. 4.41). This ultrafast magnetization is ascribed to rapid hole injection from $SrTiO_3$ into the manganite due to strong band bending of $SrTiO_3$ near the interface, while the long lifetime of the magnetization is attributed to a delay in electron injection to the manganite layer [57].

References

[1] T. Saitoh, A. E. Bocquet, T. Mizokawa, H. Namatame, A. Fujimori, M. Abbate, Y. Takeda, and M. Takano, *Phys. Rev. B* 51, 13942 (1995).

[2] Y. Okimoto, T. Katsufuji, T. Ishikawa, T. Arima, and Y. Tokura, *Phys. Rev. B* 55, 4206 (1997).

[3] M. Shi, M. C. Falub, P. R. Willmott, J. Krempasky, R. Herger, L. Patthey, K. Hricovini, C. V. Falub, and M. Schneider, *J. Phys.: Condens. Matter* 20, 222001 (2008).

[4] M. Rini, Y. Zhu, S. Wall, R. I. Tobey, H. Ehrke, T. Garl, J. W. Freeland, Y. Tomioka, Y. Tokura, A. Cavalleri, and R. W. Schoenlein, *Phys. Rev. B* 80, 155113 (2009).

[5] V. Kiryukhin, D. Casa, J. P. Hill, B. Keimer, A. Vigiante, Y. Tomioka, and Y. Tokura, *Nature* 386, 813 (1997).

[6] K. Miyano, T. Tanaka, Y. Tomioka, and Y. Tokura, *Phys. Rev. Lett.* 78, 4257 (1997).

[7] M. Fiebig, K. Miyano, Y. Tomioka, and Y. Tokura, *Science* 280, 1925 (1998).

[8] M. Rini, R. Tobey, N. Dean, J. Itatani, Y. Tomioka, Y. Tokura, R. W. Schoenlein, and A. Cavalleri, *Nature* 449, 72 (2007).

[9] H. Oshima, M. Nakamura, and K. Miyano, *Phys. Rev. B* 63, 075111 (2001).

[10] J. M. Dai, W. H. Song, S. G. Wang, S. L. Ye, K. Y. Wang, J. J. Du, and Y. P. Sun, *J. Appl. Phys.* 90, 3118 (2001).

[11] J. M. Dai, W. H. Song, J. J. Du, J. N. Wang, and Y. P. Sun, *Phy. Rev. B* 67, 144405 (2003).

[12] Y. Tokura, *J. Phys. Soc. Jpn.* 75, 011001 (2006).

[13] N. Takubo, Y. Ogimoto, M. Nakamura, H. Tamaru, M. Izumi, and K. Miyano, *Phys. Rev. Lett.* 95, 017404 (2005).

[14] N. Takubo, I. Onishi, K. Takubo, T. Mizokawa, and K. Miyano, *Phys. Rev. Lett.* 101, 177403 (2008).

[15] Y. Tomioka, A. Asamitsu, Y. Moritomo, H. Kuwahara, and Y. Tokura, *Phys. Rev. Lett.* 74, 5108 (1995).

[16] L. Hu, Y. P. Sun, B. Wang, Z. G. Sheng, X. Luo, X. B. Zhu, Z. R. Yang, W. H. Song, J. M. Dai, Z. Z. Yin and W. B. Wu, *J. Appl. Phys.* 106, 083903 (2009).

[17] S. Chaudhuri, N. K. Pandey, S. Saini, and R. C. Budhani, *J. Phys.: Condens. Matter* 22, 275502 (2010).

[18] L. Hu, Z. G. Sheng, X. Luo, Y. Liu, Z. H. Huang, W. H. Song, and Y. P. Sun, *Phys. Scr.* 87, 055701 (2013).

[19] K. Nasu, *Photoinduced Phase Transitions* (World Scientific, 2004).

[20] H. Ichikawa, S. Nozawa, T. Sato, A. Tomita, K. Ichiyanagi, M. Chollet, L. Guerin, N. Dean, A. Cavalleri, S. Adachi, T. Arima, H. Sawa, Y. Ogimoto, M. Nakamura, R. Tamaki, K. Miyano, and S. Koshihara, *Nat. Mater.* 10, 101 (2011).

[21] V. N. Smolyaninova, M. Rajeswari, R. Kennedy, M. Overby, S. E. Lofland, L. Z. Chen and R. L. Greene, *Appl. Phys. Lett.* 86, 071922 (2005).

[22] V. I. Kudinov, I. L. Chaplygin, A. I. Kirilyuk, N. M. Kreines, R. Laiho, E. Lahderanta, and C. Ayache, *Phy. Rev. B* 47, 9017 (1993).

[23] G. Nieva, E. Osquiguil, J. Guimpel, M. Maenhoudt, B. Wuyts, Y. Bruynseraede, M.B. Maple, and I.K. Schuller, *Appl. Phys. Lett.* 60, 2159 (1992).

[24] R. Cauro, A. Gilabert, J. P. Contour, R. Lyonnet, M.-G. Medici, J.-C. Grenet, C. Leighton, and Ivan K. Schuller, *Phy. Rev. B* 63, 174423 (2001).

[25] Z. G. Sheng, Y. P. Sun, J. M. Dai, X. B. Zhu, and W. H. Song, *Appl. Phys. Lett.* 89, 082503 (2006).

[26] X. Yuan, Z. J. Yan, Y. B. Xu, G. M. Gao, K. X. Jin, and C. L. Chen, *Appl. Phys. Lett.* 90, 224105 (2007).

[27] E. Beyreuther, A. Thiessen, S. Grafstrom, L. M. Eng, M. C. Dekker, and K. Dorr, *Phys. Rev. B* 80, 075106 (2009).

[28] V. N. Smolyaninova, G. Yong, Rajeswari M. Kolagani, K. Karki, and B. Hofmann, *J. Appl. Phys.* 106, 043902 (2009).

[29] L. Hu, Z. G. Sheng, Y. N. Huang, W. H. Song, and Y. P. Sun, *Appl. Phys. Lett.* 101, 042413 (2012).

[30] Y. G. Zhao, J. J. Li, R. Shreekala, H. D. Drew, C. L. Chen, W. L. Cao, C. H. Lee, M. Rajeswari, S. B. Ogale, R. Ramesh, G. Baskaran, and T. Venkatesan, *Phys. Rev. Lett.* 81, 1310 (1998).

[31] X. J. Liu, Y. Moritomo, A. Machida, A. Nakamura, H. Tanaka, and T. Kawai, *Phys. Rev. B* 63, 115105 (2001).

[32] V. Moshnyaga, A. Giske, and K. Samwer, E. Mishina, T. Tamura, and S. Nakabayashi, A. Belenchuk, O. Shapoval, and L. Kulyuk, *J. Appl. Phys.* 95, 7360 (2004).

[33] L. Hu, Y. P. Sun, B. Wang, X. Luo, Z. G. Sheng, X. B. Zhu, W. H. Song, Z. R. Yang, and J. M. Dai, *Chin. Phys. Lett.* 27, 097504 (2010).

[34] V. N. Smolyaninova, G. Yong, R. M. Kolagani, A. Biswas, H. K. Ermer, K. Wang, and A. Piazza, *Appl. Phys. Lett.* 99, 222507 (2011).

[35] E. J. Guo, L. Wang, H. B. Lu, H. Z. Guo, M. He, K. J. Jin, G. Z. Yang, and J. Gao, *Appl. Phys. Lett.* 100, 061902 (2012).

[36] K. Matsuda, A. Machida, Y. Moritomo, and A. Nakamura, *Phys. Rev. B* 58, 4203 (1998).

[37] M. Baran, S. L. Gnatchenko, O. Yu. Gorbenko, A. R. Kaul, R. Szymczak, and H. Szymczak, *Phys. Rev. B* 60, 9244 (1999).

[38] H. Huhtinen, R. Laiho, E. Lahderanta, L. S. Vlasenko, M. P. Vlasenko, and V. S. Zakhvalinskii, *Phys. Rev. B* 62, 11614 (2000).

[39] Y. Okimoto, Y. Ogimoto, M. Matsubara, Y. Tomioka, T. Kageyama, T. Hasegawa, H. Koinuma, M. Kawasaki and Y. Tokura, *Appl. Phys. Lett.* 80, 1031 (2002).

[40] H. Katsu, H. Tanaka, and T. Kawai, *Appl. Phys. Lett.* 76, 3245 (2000).

[41] J. R. Sun, C. M. Xiong, B. G. Shen, P. Y. Wang, and Y. X. Weng, *Appl. Phys. Lett.* 84, 2611 (2004).

[42] J. R. Sun, C. H. Lai and H. K. Wong, *Appl. Phys. Lett.* 85, 37 (2004).

[43] J. R. Sun, S. Y. Zhang, B. G. Shen, and H. K. Wong, *Appl. Phys. Lett.* 86, 053503 (2005).

[44] J. R. Sun, B. G. Shen, Z. G. Sheng, and Y. P. Sun, *Appl. Phys. Lett.* 85, 3375 (2004).

[45] H. B. Lu, K. J. Jin, Y. H. Huang, M. He, K. Zhao, B. L. Cheng, Z. H. Chen, Y. L. Zhou, S. Y. Dai, and G. Z. Yang, *Appl. Phys. Lett.* 86, 241915 (2005).

[46] Z. G. Sheng, B. C. Zhao, W. H. Song, Y. P. Sun, J. R. Sun, and B. G. Shen, *Appl. Phys. Lett.* 87, 242501 (2005).

[47] C. Luo, K. X. Jin, C. L. Chen, and T. Wu, *Appl. Phys. Lett.* 103, 212401 (2013).

[48] Z. G. Sheng, M. Nakamura, W. Koshibae, T. Makino, Y. Tokura, and M. Kawasaki, *Nat. Commun.* 5, 4584 (2014).

[49] H. Yamada, Y. Ogawa, Y. Ishii, H. Sato, M. Kawasaki, H. Akoh, Y. Tokura, *Science* 305, 646 (2004).

[50] W. M. Lu, J. R. Sun, Y. Z. Chen, and B. G. Shen, *Appl. Phys. Lett.* 94, 152514 (2009).

[51] W. W. Gao, A. D. Wei, J. R. Sun, D. S. Shang, J. Wang, T. Y. Zhao, and B. G. Shen, *Appl. Phys. Lett.* 96, 262503 (2010).

[52] M. Minohara, R. Yasuhara, H. Kumigashira, and M. Oshima, *Phys. Rev. B* 81, 235322 (2010).

[53] A. D. Wei, J. R. Sun, W. M. Lu, and B. G. Shen, *Appl. Phys. Lett.* 95, 052502 (2009).

[54] W. W. Gao, L. Hu, Y. P. Sun, J. R. Sun, J. Shen, R. J. Chen, Y. F. Chen, and B. G. Shen, *J. Appl. Phys.* 113, 17D716 (2013).

[55] A. Gilabert, A. Plecenik, K. Frohlich, S. Gazi, M. Pripko, Z. Mozolova, D. Machajdik, S. Benacka, M. G. Medici, M. Grajcar, and P. Kus, *Appl. Phys. Lett.* 78, 1712 (2001).

[56] Z. G. Sheng, Y. P. Sun, J. M. Dai, X. B. Zhu, W. H. Song, and Z. R. Yang, *J. Phys. D: Appl. Phys.* 42, 085002 (2009).

[57] H. Yada, M. Matsubara, H. Yamada, A. Sawa, H. Matsuzaki, and H. Okamoto, *Phys. Rev. B* 83, 165408 (2011).

In: Manganite-Based Thin Films ...
Editors: Z. Sheng, L. Hu and Y. Sun
ISBN: 978-1-63482-452-1

Chapter 5

MAGNETO-TUNABILITY OF IN MANGANITE FILMS AND HETEROJUNCTIONS

Zhi Gao Sheng*
High Magnetic Field Laboratory, Chinese Academy of Science,
Hefei, Anhui, China

Abstract

Magneto-tunability is one of the fantastic features of correlated manganite compounds. The realization of correlated interface in manganite-based heterojunctions, thus, is expected to be sensitive to the external magnetic field, which may create new functionalities of correlated electron devices. In this chapter, after a brief overview of the magnetic field effect in manganite, a review of the magneto-tunable effect of manganite-based heterojunctions is given with special attention to the manganite/Nb:$SrTiO_3$. Moreover, the magnetic field influence on photovoltaic effect in those junctions is also discussed.

PACS: 74.20.Mn, 74.25.Dw

Keywords: magnetic field, manganite, heterojunction, photovoltaic effect

1. Introduction of Magnetic Field Effect in Manganite

With application of a magnetic field, the electrical resistance of a material will change, which is so-called as magnetoresistance (MR) property. There are sev-

*E-mail address: zhigaosheng@hmfl.ac.cn

eral kinds of magnetoresistance existing in different materials, including ordinary magnetoresistance, anisotropic magnetoresistance, giant magnetoresistance, and colossal magnetoresistance [1].

Ordinary Magnetoresistance (OMR). For non-magnetic metals, MR effects at low fields are very weak, although the effect can become quite strong for high fields. The change in resistivity is positive for both magnetic field parallel and transverse to the current direction. In ordinary magnetoresistance, Lorenz force plays a very important role in the charge carriers and the increases of the resistance in an applied magnetic field. Depending on the structure of the electron orbitals at the Fermi surface, there are three distinct cases of ordinary magnetoresistance. 1) In metals with closed Fermi surfaces, the electrons are constrained to their orbit in k space and the effect of the magnetic field is to increase the cyclotron frequency of the electron in its closed orbit. Metals which exhibit this behavior include In, Al, Na and Li. 2) For metals with equal numbers of electrons and holes, the magnetoresistance increases with H up to the highest fields measured, and is independent of crystallographic orientation. Examples of metals displaying this behavior are Bi, Sb, W, and Mo. 3) Metals that contain Fermi surfaces with open orbits in some crystallographic directions will exhibit large magnetoresistance for fields applied in those directions, whereas the resistance will saturate in other directions, where the orbits are closed. This behavior will be found in Cu, Ag, Au, Mg, Zn, Cd, Ga, TI, Sn, Pb, and Pt.

Anisotropic Magnetoresistance (AMR). In ferromagnetic metals and alloys, MR effects on the order of $\Delta\rho/\rho$ of 2% are obtained in low fields. In contrast to OMR, the effect is anisotropic, where $\Delta\rho_{||}$ increases with field, and $\Delta\rho_{\perp}$ decreases with field. The physical origin of the magnetoresistance effect lies in spin orbit coupling. The electron cloud of each nucleus deforms slightly as the direction of the magnetization rotates, and this deformation changes the amount of scattering undergone by the conduction electrons when traversing the lattice. A heuristic explanation is that the magnetization direction rotates the closed orbit orientation with respect to the current direction. If the field and magnetization are oriented transverse to the current, then the electronic orbits are in the plane of the current, and there is a small cross-section for scattering, giving a low resistance state. Conversely for fields applied parallel to the current, the electronic orbits are oriented perpendicular to the current, and the cross-section for scattering is increased, giving a high resistance state.

Giant Magnetoresistance (GMR). It is a quantum mechanical magnetoresistance effect observed in thin-film structures composed of alternating ferromagnetic and non-magnetic conductive layers. The effect is observed as a significant change in the electrical resistance depending on whether the magnetization of adjacent ferromagnetic layers are in a parallel or an antiparallel alignment, which was discovered in 1988 by Baibich et al. in antiferromagnetically coupled multilayers of Fe/Cr. The overall resistance is relatively low for parallel alignment and relatively high for antiparallel alignment. The magnetization direction can be controlled, for example, by applying an external magnetic field. The effect is based on the dependence of electron scattering on the spin orientation. Magnetoresistance effects of up to -50% were observed at low temperatures. This effect was subsequently found to occur in a number of multilayer magnetic film systems. The main application of GMR is magnetic field sensors, which are used to read data in hard disk drives, biosensors, microelectromechanical systems (MEMS) and other devices. GMR multilayer structures are also used in magnetoresistive random-access memory (MRAM) as cells that store one bit of information. In 2007,the Nobel Prize in Physics was awarded to Albert Fert and Peter Grünberg for the discovery of GMR.

Colossal Magnetoresistance (CMR). It has been known since the early 1950's that manganese oxides with perovskite structures, when doped, become ferromagnetic metals, and exhibit remarkable magnetoresistive properties. The largest effects have been observed for x=0.33 in $La_{1-x}A_xMnO_3$ (A = Ca, Sr). The term colossal has arisen from the huge effects observed, on the order of $\Delta R/R(H)$ = 125,000%. If normalized to the zero field values, the resistance changes by 99.90%. Zener explained magnetism in these materials via the double exchange (DE) mechanism [2]. He assumed that the only way charge transport between Mn^{4+} and Mn^{3+} can happen is via the simultaneous hopping of an e_g-electron from Mn^{3+} to the connecting O^{2-} and from the O^{2-} to the Mn^{4+} e_g band; hence the term double exchange [3]. Ferromagnetism is then induced via this hopping by the very large Hund's rule coupling between the Mn eg and t_{2g} electrons, resulting from the high spin state of the manganese d-electrons (see Chapter **1**).

In this chapter, we will introduce the magnetoresistance in manganite films and heterojunctions. And then, the magneto-tunability of photon-induced effect in manganite based heterojunctions, including photovoltaic effect, will be presented.

2. Magnetoresistance in Manganite Films and Heterojunctions

2.1. Crossover of Magnetoresistance in Manganite-based Heterojunctions

In the bulk and thin films of doped manganese oxides, the CMR properties have been intensively studied since the discovery of such effect. From the beginning of this century, much attention has been paid to the fabrication of artificially designed structures to verify new device concepts based on oxide materials such as magnetic tunnel junctions [4], *p-n* junctions [5], and superlattices [6]. Very high negative MR ratios in these artificially designed structures have been reported. For example, a trilayer tunneling junction $La_{0.67}Sr_{0.33}MnO_3/SrTiO_3/$ $La_{0.67}Sr_{0.33}MnO_3$ even shows a MR as high as $\sim$ -1800% at low temperatures [7]. And a MR (defined as $\Delta\rho/\rho(0)$) as high as -55% in a $La_{2/3}Ca_{1/3}MnO_3/Pr_{2/3}Ca_{1/3}MnO_3$ superlattice at 219 K in 0.05 T has also been reported [6]. The reason for negative MR is not very clear at present, though one argue that it may originate from the negative MR of the manganites.

However, a few perovskite oxide materials with positive CMR have also been found. Ghosh et al. observed a $\Delta\rho/\rho(0)$ of 25% in a $Fe_3O_4/SrTiO_3/La_{0.7}Sr_{0.3}MnO_3$ magnetic tunneling junction (MTJ) at 80 K in a 4 T field [8]. Mitra et al. discovered a large positive MR in $La_{0.7}Ca_{0.3}MnO_3$ $/SrTiO_3$ $/La_{0.7}Ce_{0.3}MnO_3$ MTJ at 48 K in 2 T [4].2 Positive MR ratios of 1800% in a $Pr_{0.6}Ca_{0.4}MnO_3$/Nb-doped-$SrTiO_3$ *p-n* junction at 200 K in 5 T and 1100% in weak $La_{0.67}Ca_{0.33}MnO_{3-\delta}$/Nb-doped $SrTiO_3$ *p-n* junction at 120 K in 1.7T were also reported in our previous work [10, 11]. Although the magnetically tunable feature of manganite-based heterojunctions, similar to manganites, has been verified, the factors affecting the sign of MR have not been reported yet. To clarify this issue in detail, Sheng *et al.* studied the $La_{0.7}Ce_{0.3}MnO_3$ (LCEM) based heterojunctions carefully and a crossover from negative to positive magnetoresistance was observed in the heterojunction composed of a LCEM film and a 0.5 wt% Nb-doped $SrTiO_3$ (STON) substrate [12].

In this study, the LCEM/STON heterojunction was fabricated by growing a LCEM film on 3×5 mm^2 STON crystal with the (001) orientation using laser ablation. The film was $\sim$500 Å in thickness. For improving crystal quality and keeping the oxygen stoichiometry, the sample was maintained at 750 ^oC for 20 min in flowing O_2 gas after deposition. X-ray diffraction analysis showed that

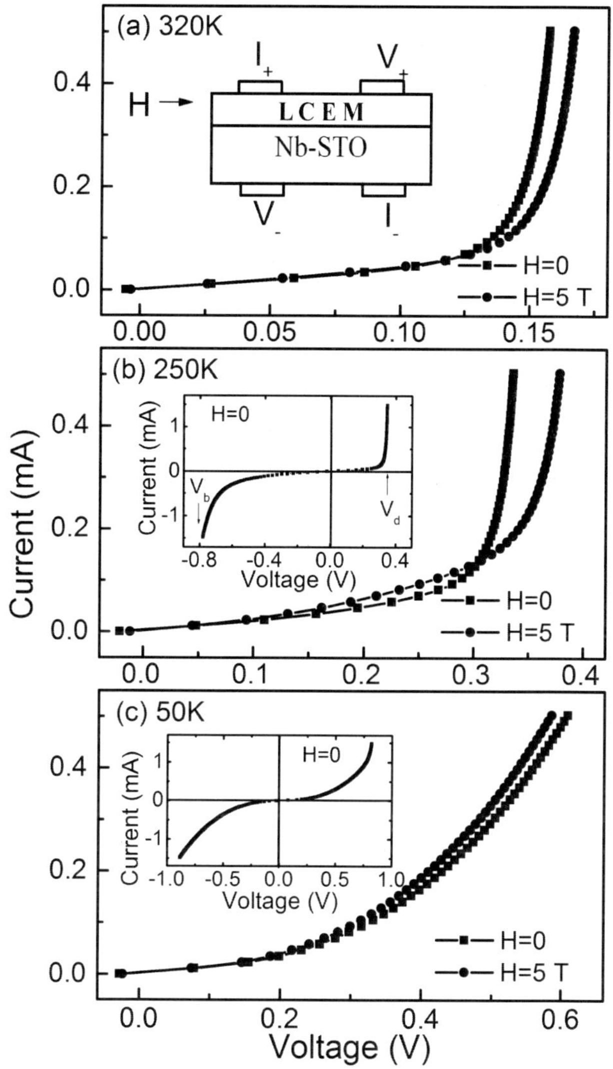

Figure 5.1. $I - V$ characteristics of LCEM/STON heterojucntion measured by tuning bias currents. The inset of (a) shows a schematic illustration of electrode setting. The insets of (b) and (c) present I-V curves at temperature of 250 K and 50 K, respectively (Reproduced with permission from Reference [12]. Copyright 2005, AIP Publishing LLC).

the prepared film was single phase and (110) oriented.

The current-voltage ($I - V$) behavior of the LCEM/STON was measured by tuning bias current under applied magnetic fields varying between 0 and 5 T and a temperature range of 5~320 K by a Quantum Design Physical Property Measurement System (PPMS). To minimize the heating effect at low temperature, the measuring current was kept below 1.5 mA. The typical results are presented in Fig. 5.1. It indicates that the $I - V$ curve shows good rectifying behavior at higher temperatures, and nearly asymmetric behavior occurs at temperature below 50 K. However, the $I - V$ curves still exhibit nonlinearity at lower temperatures. These results imply a weakening of the rectifying ability of the LCEM/STON junction with decreasing temperatures. Compared with the previously reported manganite heterojunction, LCEM/STON exhibits a rather complex $I - V$ relation with magnetic fields. In addition to the two current rushes at V_b (breakdown voltage) and V_d (diffusion voltage), respectively, which are typical features of heterojunctions, the $I - V$ curve under zero and 5 T magnetic fields shows a significant magnetoresistance (defined as $MR = [R_H - R_0]/R_0 \times 100\%$) within whole measured temperature range. And the most remarkable observation of the present work is that there exists a crossover from negative to positive MR in the LCEM/STON heterojunction. There are two obvious features in the variation of MR. First, the MR is temperature dependent. As shown in Fig. 5.1(a), the $I - V$ curves under zero and 5 T magnetic fields recorded at T=320 K show a significant positive MR. However, the sign of MR change from positive to negative when the temperature is decreased below a critical value, as shown in Fig. 5.1(c). Second, the MR is electric field bias dependent. The typical results at 250 K are shown in Fig. 5.1(b). It indicates that the sign of MR is negative under a lower bias voltage. However, with increasing applied voltage, the sign of MR changes to positive. This kind of phenomenon is observed in both positive and negative current bias regions. All these effects show magnetically tunable features in the LCEM/STON heterojunction.

In order to explore the variation of MR at different temperatures further, MR of LCEM/STON junction under a field of 5 T and at selected bias currents I=-1.5 mA, 1.5 mA, and 0 (obtained by extrapolating the results to the zero-current limit) has been studied, and the results are shown in Fig. 5.2. At a zero bias current the MR is negative within the whole measured temperature range and has a maximum value at the temperature of 275 K. While, at non-zero current (I= 1.5 mA), the MR is quite different from the former. It changes from negative

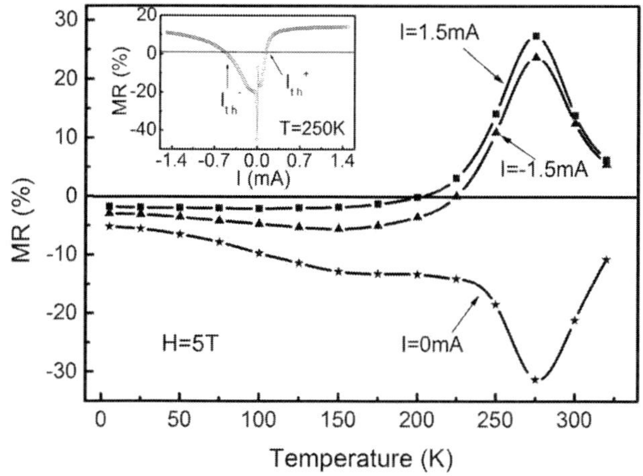

Figure 5.2. Magnetoresistance (MR) of the LCEM/STON junction measured under a field of 5 T and at selected current I=-1.5,0,1.5 mA. The inset displays the MR of LCEM/STON junction as a function of bias currents under a field of 5 T and at the selected temperature of 250 K (Reproduced with permission from Reference [12]. Copyright 2005, AIP Publishing LLC).

to positive values as the temperature increases. It is obvious that the temperature dependence of the sign of MR of the LCEM/STON junction is quite different from that of the manganites themselves. One point deserving special attention is that the maximum for both positive and negative MR occurs at 275 K, which is equal to the Curie temperature of LCEM. These results reveal that there exists a close relation between electric transport and magnetization of LCEM/STON heterojunction.

The inset of Fig. 5.2 presents the MR as a function of applied currents at 250 K. It displays obviously that the applied current also causes a sign change of the MR. The MR increases with increasing bias currents and there exists a critical current I_{th} at which the sign of MR experiences a negative to positive change. The critical electric current I_{th} decreases with increasing temperatures and the value of I_{th} is asymmetric with respect to the direction of the applied current. For manganite-based heterojunctions, we know that the relation between MR and applied magnetic fields is very important for both physics and their potential applications as magnetic random access memories (MRAMS) or sensors. Fig.

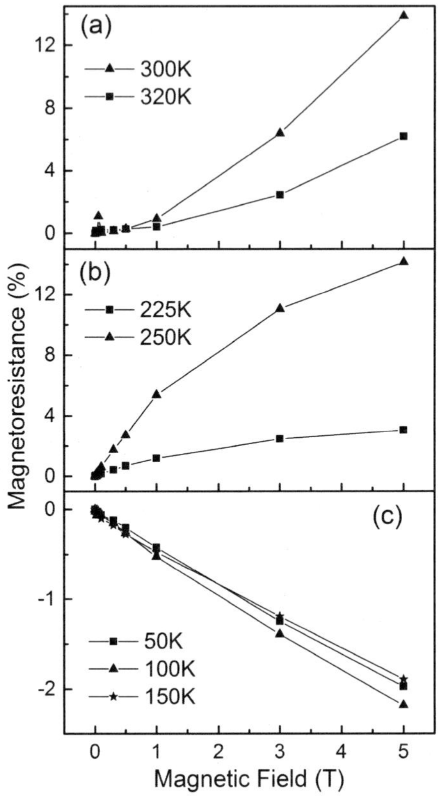

Figure 5.3. Magnetoresistance of the LCEM/STON junction as a function of magnetic field at the selected current I=1.5 mA (Reproduced with permission from Reference [12]. Copyright 2005, AIP Publishing LLC).

5.3 presents MR as a function of magnetic fields at different temperatures and a given bias current of 1.5 mA. It shows obviously that MR has a simple relation with external magnetic fields when MR has a positive sign,

$$MR \propto \alpha H^{\beta} \quad (1)$$

where α and β are coefficients. The results, as shown in Fig. 5.3(a) and (b), indicate that the coefficients β are quite different when temperature is swept from above T_C to below T_C. That is to say, β is larger than 1 (1.74 for 320 K and 1.56 for 300 K) when temperature is higher than T_C, while smaller than 1

(0.68 for 250 K and 0.65 for 225 K) when temperature is lower than T_C. With deceasing temperature, the sign of MR changes to negative. Interestingly, the results shown in Fig. 5.3(c) reveal the monotonic increase of MR with magnetic fields. The relation between MR and magnetic field can be described by

$$MR \propto \gamma H \tag{2}$$

where γ is a coefficient. It reveals that there exist different transport properties of LCEM/STON junction in different temperature regions. A thorough understanding of the present observation requires a full knowledge of magnetic properties of LCEM/STON heterojunction.

It is well known that the breakdown voltage (V_b) and diffusion voltage (V_d) are two very important parameters describing heterojunctions, and their variation behavior enables us to investigate the transport mechanism of heterojunctions. Fig. 5.4 shows V_d and V_b of LCEM/STON heterojunction as a function of temperature. With increasing temperatures, the V_d decreases non-linearly and changes sharply in the vicinity of Curie temperature of LCEM film. According to the semiconductor theory and temperature dependence of V_d, it can be deduced that the energy gap between LCEM and STON decreases with increasing temperatures, so do the built-in field, junction resistance [13]. This observation is consistent with the half-metal character of the manganites. In the manganites, spin-up and spin-down of eg bands act as valence and conduction bands, respectively, and the increase in temperature could result in a diminution of band gap due to the reduction of ferromagnetic order. This may be the reason why the increasing temperature gives rise to the decrease of V_d.

As to V_b(T), it is quite different from V_d(T) and has a local maximum appearing near 175 K, i.e., V_b increases with increasing temperatures below 175 K, and decreases with increasing temperatures above 175 K. The different temperature dependence of V_b below and above 175 K may originate from different reverse transport mechanism. Based on semiconductor theory, since the energy gap of LCEM/STON decreases with increasing temperatures as evidenced by V_d, the V_b in the heterojunction due to the tunneling effect has a negative temperature coefficient, that is, the V_b decreases with increasing temperatures. This kind of temperature dependence is quite different from the ionization mechanism, in which the V_b has a positive temperature coefficient. In our experiments, carrier transport via the tunneling mechanism may be dominant when the temperature is above 175 K. Similar phenomenon was also observed in the other manganite-STON heterojunctions [10], which indicates reverse tunneling transport mecha-

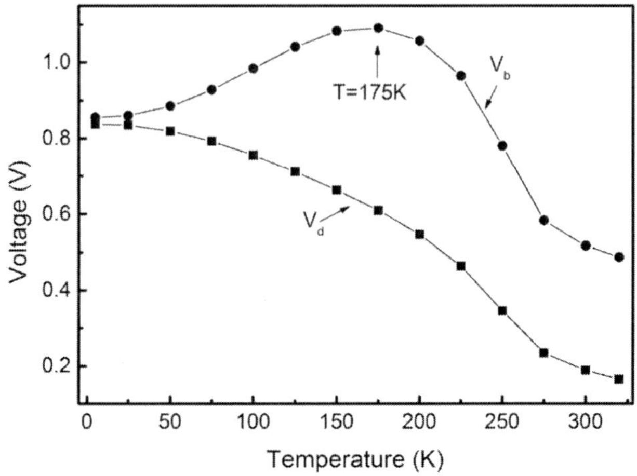

Figure 5.4. Temperature dependence of diffusion voltage (V_d) and breakdown voltage (V_b) of the LCEM/STON junction (Reproduced with permission from Reference [12]. Copyright 2005, AIP Publishing LLC).

nism could be a common feature in manganite-STON heterojunctions.

The temperature and current bias dependence of the MR are also observed in a magnetic tunneling junction by Mitra et al [4]. They found that the MR in La0.7Ce0.3MnO$_3$ /SrTiO$_3$/La0.7Ca0.3MnO$_3$ MTJ is negative at small bias currents, and turns to positive at higher bias currents at 100 K. It is interesting that the positive MR ratios in our experiments only exist at high temperatures that just lie in temperature range that tunneling dominates the electric transport of the heterojunction. Therefore, it seems reasonable to attribute the appearance of positive MR in the LCEM/STON heterojunction to the tunneling transport mechanism. Supposing an electric current tunneling from LCEM into paramagnetic metal STON, it should be considered that there exist two channels (spin up $I_\uparrow$ and spin down $I_\downarrow$) for electrons to tunnel, as shown in Fig. 5.5. The ratio $I_\uparrow/I_\downarrow$ is proportional to the spin polarization of LCEM. At higher temperatures, the ratio $I_\uparrow/I_\downarrow$ is approximate to 1 because the spin polarization of LCEM is lower. If an external magnetic field is applied to the junction, the disorder of spin is suppressed, thus the spin polarization and the ratio $I_\uparrow/I_\downarrow$ increase, as shown in Fig. 5.5(b). It means that one channel for tunneling is shut down by the applied magnetic field. It results in the increase of the resistance of junction

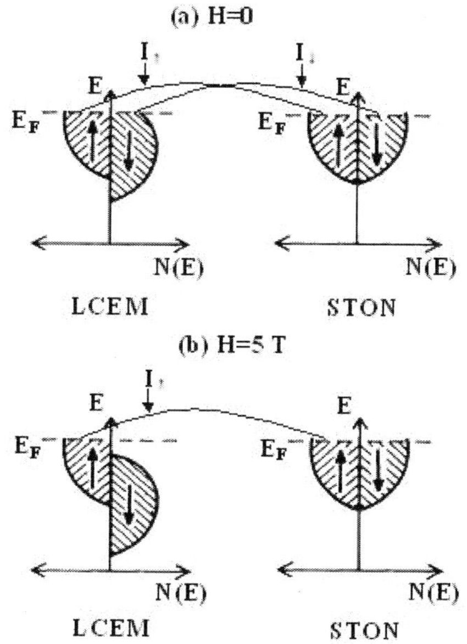

Figure 5.5. Schematic diagrams for the band structure of LCEM and STON at (a) H=0 and (b) H=5 T (Reproduced with permission from Reference [12]. Copyright 2005, AIP Publishing LLC).

and a positive MR. In the vicinity of T_C, the variation of electric and magnetic properties of LCEM caused by the applied magnetic field is significant, so is the MR. This feature is proved by our experiments.

2.2. Coaction of Electric and Magnetic Field Effect on Manganite Films

Manganites show a wide variety of phases associated with an interplay between spin, charge, orbital, and lattice degrees of freedom [15]. Application of a slight external field such as magnetic field, electric field, photon or pressure to a manganite which is located at the phase boundary will get huge and quick response. Among them, the electric field induced polarization effect on manganites has attracted much attention for its possible application in new field-effect devices. Pallecchi et al. observed a shift of the metal-insulator transition temperature as high as 43 K and a resistivity modulation up to 250% in $La_{0.7}Sr_{0.3}MnO_3$ side-gate channels [16]. Ahn et al. found that the transport characteristics of LSMO vary sharply using the polarization field from a ferroelectric oxide, $Pb(Zr,Ti)O_3$ [17]. They showed us the clear depletion of holes in manganite films due to the polarization effect. A large electroresistance $\sim 76\%$ at 40kV/cm was also found in $(LaCa)MnO_3$ (LCMO) films with $Pb(Zr,Ti)O_3$ gate and was explained by the polarization effect on percolative phase separation [18].

Besides the electric induced polarization effect, lattice deformations caused by some kinds of strain is another important factor to control the electronic states in manganites because of the strong electron-lattice coupling [19]. The commonly applied method for the study of the lattice strain effect is the growth of thin films on substrates with a certain lattice mismatch. The substrate-induced lattice strain, which could change the strength of the double-exchange (DE) interaction and Jahn-Teller ($J - T$) electron-lattice coupling via modifying Mn-O bond lengths and/or Mn-O-Mn bond angles, is one of the most important factors that strongly affect the properties of manganite thin films, such as Curie temperature (T_C), magnetroresistance, and phase separation. For example, the metal-insulator transition temperature T_{MI} of $La_{0.7}Sr_{0.3}MnO_3$ (LSMO) can be reduced from 365 K for unstrained films/bulk to 270 K in 16 nm thick LSMO/$LaAlO_3$(001) films strained compressively [20]. Millis et al. proved that the ferromagnetic T_C is extremely sensitive to biaxial strain [21]. Gao et al. found that the transition temperature is correlated with the strain state in epitaxial LCMO films [22]. Generally, it is difficult to modulate the strain effect in manganite films continuously and reversely by using conventional substrates. For overcoming this obstacle, some experiments, in which a ferroelectric crystal with electrostrictive or piezoelectric properties were used as substrate, have been done [23, 24, 25]. The strain-induced shifts of the T_C and enhancement of magnetization of $La_{0.7}A_{0.3}MnO_3$ films have been found [24]. Zheng et al. also reported the charge-ordered state and paramagnetic state of $Pr_{0.5}Ca_{0.5}MnO_3$ film can be modulated continuously by applying dc or ac electric fields across ferroelectric crystal [25]. As a result of using ferroelectric crystal, the field induced polarization effect should also be considered except field induced strain effect. Hence, thorough knowledge of the co-action of these two factors benefits not only to clarification of the electric induced effects in manganites but also to the designing of artificial materials.

X-ray diffraction measurements ($\theta - 2\theta$) performed in LCMO/PMN-PT exhibit the response of the c lattice parameter of PMN-PT(001) crystal under applied electric field [26]. The typical results are shown in Fig. 5.6. It is found that the c constant depends in a nearly linear way on the bias electric field.

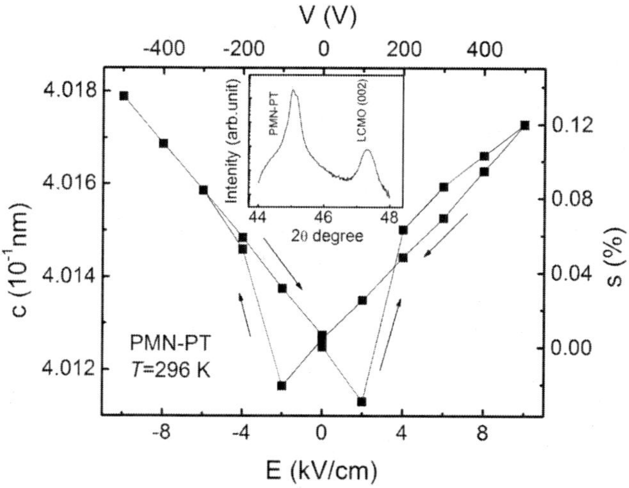

Figure 5.6. The variation of out-of-plane lattice constant c of a PMN-PT(001) crystal and strain s under an electric field E along the c axis. The inset shows the x-ray diffraction pattern ($\theta - 2\theta$) at angles in the vicinity of the pesudocubic (002) reflection for LCMO/PMN-PT (Reprinted with permission from Reference [26], Copyright (2009) by the American Physical Society).

It increases by about 0.12% with an electric field of E=10 kV/cm (V_{bias}=500 V). The relation between lattice strain s and E indicates a converse piezoelectric nature of the crystal. More precisely, it can be found that the slope of strain decreases slightly with increasing voltage. Cycling the field from positive to negative values, a butterfly-like hysteresis loop with coercive fields of 2 kV/cm(V_{bias}=100 V) is observed. Assuming approximate volume conservation, 0.12% of out-of-plane expansion would be accompanied by 0.06% of contraction in both in-plane directions of the PMN-PT(001) single crystal [24]. If we consider that the in-plane of LCMO film is compressed due to the contraction strain coming from PMN-PT substrate, the in-plane distortion ration $\varepsilon_{xx} = \varepsilon_{yy}$ should be -0.06% and the out-plane distortion ratio ε_{zz} is about +0.043% using ε_{xx}=-$2\nu/(1-\nu)\varepsilon_{zz}$ and Poission coefficient ν ~0.41 [27].

The temperature dependence of the resistance of LCMO film is plotted in right inset of Figure **2.9**(b). A metal-insulator transition at T_{MI}=201K was observed. The electric field control of electrical properties in LCMO films was measured in the field-effect-transistor (FET) configuration as shown in the left inset of Figure **2.9**(b). The resistance of LCMO was measured through source and drain electrodes and the bias electric voltages were applied to the LCMO/PMN-PT structure through the gate and drain. A resistor of 4.7 $M\Omega$ was connected in series with the gate in order to protect the electrical meters in case a dielectric breakdown took place in the PMN-PT crystal. Results of transport measurements of LCMO film under bias electric field are shown in Figure **2.9**. The resistance R at room temperature drops by about -3.61% ($\Delta R/R = [R(E) - R(0)]/R(0)$) at E=-8 kV/cm due to the electric induced in-plane strain in the film. (The positive bias is defined that when the electrode on PMN-PT gate is positive while the electrode on LCMO is negative.) The change of resistance with variation of V is not linear. It is similar to the strain coefficient s shown in Fig. 5.6 that a butterfly-like hysteresis loop with coercive fields of 2 kV/cm is also observed when we cycle the electric field from positive to negative values. On the basis of the theory of strain effect on manganite films, it is easy to understand this butterfly-like behavior as follow. In the simplest scenario, two aspects caused by strain in manganite films can be envisaged: a uniform compression will tend to increase the bare electron hopping amplitude and thereby reduce the importance of the electron-lattice coupling, at the same time a biaxial distortion will also compete with the $J - T$ distortion and lattice distortion caused by tolerance factor, thus affecting the properties of manganite films especially for magnetic structure (it will be discussed later). Due to the first aspect, the enhancement of electron hopping amplitude will contribute to the decreasing of resistance as shown in Figure **2.9**(a).

As temperature decreases, an interesting phenomenon occurs as shown in Figure **2.9**(b). In the negative branch, the resistance of LCMO film decreases with increasing negative bias voltage. In the positive branch, however, the resistance increases with increasing positive bias voltage, which is quite different from that of high temperatures. As we can see from Figure **2.9**, the butterfly-like hysteresis loop changes into shuttle-like loop with decreasing of temperature. At 50 K, the $\Delta R/R$ is about +3.48% when the electric field E is +8 V/cm, while it reaches -4.34% as E=-8 V/cm. The difference reaction to the direction of bias voltage in different temperature can be clearly reflected in Fig. 5.7. The relative change of resistance $\Delta R/R$ as function of temperatures, and bias field E=+8 and -8 V/cm respectively, were depicted in this figure. It can be found that the resistance decreases in spite of the direction of bias electric field at high temperature. We named this area as region I for convenience. The change of resistance of LCMO film at low temperature, however, has a tight relation with

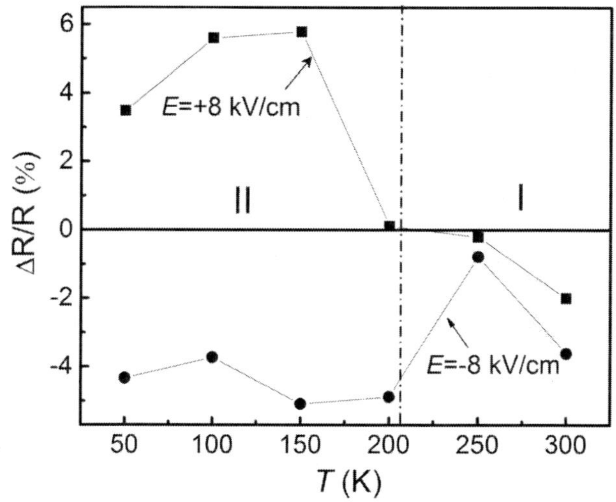

Figure 5.7. The relative change of resistance $\Delta R/R$ as function of temperatures with bias electric field E=+8 and -8 kV/cm respectively (Reprinted with permission from Reference [26], Copyright (2009) by the American Physical Society).

the direction of bias field as shown in region II. It is interesting that the transition point from region I to region II is located near T=200 K which is close to the metal-insulator transition temperature and Curie temperature (T_{MI}=201 K, T_C=203 K). These results imply that the transition from field induced butterfly-like hysteresis loop to shuttle-like behavior might has an intimate relation with the electrical and magnetic state of the LCMO films.

If only the electric field induced strain effect is considered, the shutter-like loop of relative change of resistance as function of electric field cannot be explained [21, 22]. It is difficult for the piezoelectric effect to explain the different behavior of LCMO films under different directional bias field. It should be noted that the electric field can not only induced the piezoelectric effect but also polarize the LCMO film because of using ferroelectric crystal PMN-PT. Therefore, the bias electric field applied in the FET configuration not only results in the expansion of the lattice along the direction of the electric field, but also cause the depletion or accumulation of holes in the LCMO films layer as shown in Fig. 5.8. It is known that the LCMO is a p-type materials, the majority of the

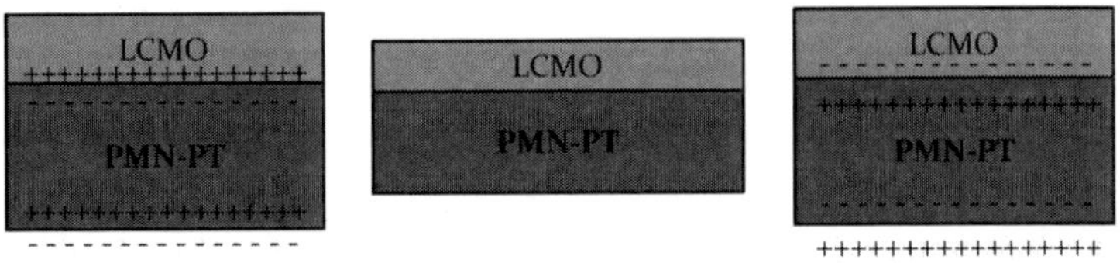

Figure 5.8. Schematic picture of co-action of electric field induced strain and polarization effect on LCMO/PMN-PT structure (Reprinted with permission from Reference [26], Copyright (2009) by the American Physical Society).

charge carries in the LCMO films are holes. The application of a positive (or negative) field to the gate will polarize the PMN-PT layer such that the electric dipole moments in the PMN-PT point upward (or downward) into the LCMO film, thereby leading to a depletion (or accumulation) of holes in the interface part of LCMO film as shown in Fig. 5.8(a) and (c). If the polarization effect caused by external bias is taken into account, the depletion of holes will happen in LCMO film within screen length with positive bias as shown in Fig. 5.8(c). However, the screen length in LCMO films is very short because of the high carrier density of the LCMO films with doping lever x=0.3. The applied electric field will be screened within one or two lattice constants of the interface, and could not affect bulk transport of the film. In our experiments, therefore, another process may happen and it allows the electric field induced effect to penetrate more deeply into the film. The possible process for the polarization effect observed in our experiments may be relevant to the phase separation of the LCMO films. Recent experimental and theoretical studies implies that the doped manganites are intrinsically inhomogeneous and that phase separation is common in these materials [29, 28]. The phase separation and percolative transport properties in LCMO films have also been proved [18]. In an electronically inhomogeneous channel, they are partly metallic and partly insulating. The application of high electric field would change the relative volume fractions of metallic and insulating regions by accumulation of charge between metallic and insulating phases, thereby causing the interface between these phases to move [17, 18]. The nature of accumulating charge, holes or electrons, will depend on field polarity and thereby control the direction of interface movement. The field effect on the transport properties, therefore, is also field polarity dependent. In our case, the polarization effect on phase separation as well as percolative transport caused by positive external bias will make the resistance increase as shown in Figure **2.9**(b). A negative voltage does the opposite, leading to the decreasing of resistance. Considering the co-action of field induced strain and polarization effect together, we can understand the behavior that the absolute value of $\Delta R/R$ under negative bias (4.34%) is larger than that of positive bias (3.48%) at T=50 K.

In the Fig. 5.7, it can be found that the relative change of R has a maximum value near the temperature of 150 K. This value is consistent with the temperature at which the resistance of LCMO films has a steeper slope as shown in inset of Figure **2.9**(b). That is to say the modulation ΔR is clearly larger at the temperature where the change of R is steeper. This feature is another evidence for the percolative transport in LCMO films with multiphase. Another point in Fig. 5.7 should be noted that the polarization effect derived from ferroelectric PMN-PT crystal is dominant at lower temperatures and negligible at higher temperature T>T_{MI}. It reveals that the effect on phase separation of LCMO film caused by PMN-PT polarization is more effective at temperatures below the T_{MI}. This point agrees with the feature that the phase separation is significant at lower temperature in manganites film [29].

To test whether there is any influence on magnetization under co-action of strain and polarization effect, the measurements of magnetization with and without external bias electric field have been done. The data were obtained by monitoring the in-plane magnetic response to a voltage applied between the film and substrate bottom. Figure **2.10** exemplifies the response of magnetization of LCMO/PMN-PT structure to external bias field measured at a selected temperature T=10 K. When a E=+9.6 V/cm bias is applied, a sizeable decreasing of magnetization is observed, and the magnetization shows a quick switch between original state and lower magnetization state corresponding to the "bias on" and "bias off". Lower magnetization state is stable for a constant bias voltage, unaffected by the repeated bias voltage on and off. The relative variation $\Delta M/M(0)$ [defined as $[M(V) - M(0)]/M(0)$] is about -8.64%. As for the negative bias field E=-9.6 V/cm, the relative variation is about -7.47% which is smaller than that of positive bias. Before the application of larger bias field, bias field E=±4 kV/cm has been added to the LCMO/PMN-PT structure in order. However, there is no obvious response as shown in the left side of Figure **2.10** and it implies that the electric field $E \leq \pm 4$ kV/cm is insufficient to modulate

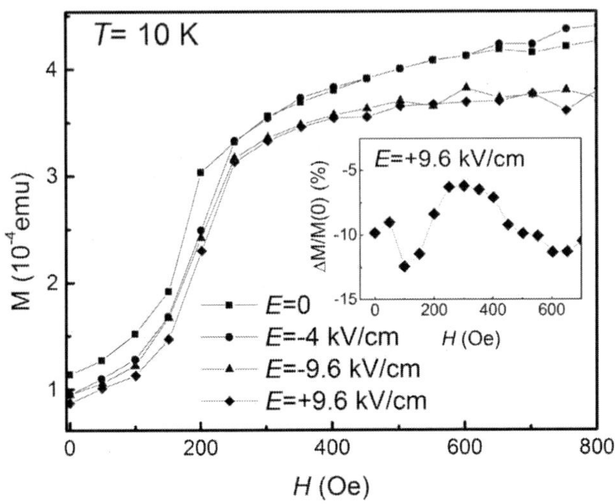

Figure 5.9. Magnetic field dependence of magnetization of LCMO/PMN-PT structure with selected bias electric field E= -4, -9.6 and +9.6 kV/cm (Reprinted with permission from Reference [26], Copyright (2009) by the American Physical Society).

the magnetic domains in LCMO films. This behavior is more clearly shown in Fig. 5.9 in which the magnetic field dependence of magnetization with selected bias field E=-4, -9.6 and +9.6 kV/cm were shown. We observed that the $M-H$ curves with E=0 and -4 kV/cm respectively are almost coincident and there is an obvious gap when bias field reaches ±9.6 kV/cm as magnetic field is more than 300 Oe. The inset of Fig. 5.9 presents the typical results of $\Delta M/M(0)$ as function of magnetic field with fixed bias field E=+9.6 kV/cm. It shows that the magnitude of $\Delta M/M(0)$ varies with magnetic field but has no clear dependence law with a fixed bias field.

It has been known that each orbit of Mn ion in manganite films has different anisotropy of the wave function and it is coupled to the displacement of the O atoms surrounding the transition-metal ion which has tight relation with strain. Moreover, the epitaxial strain could induce anisotropic hopping between orbits to cause orbit ordering at the interface. By varying the strain condition the orbital ordering changes which in turn changes the magnetic ordering at the interface. For example, the energy of $d_{x^2}-d_{y^2}$ orbit becomes higher than $d_{3z^2-r^2}$

orbit in Mn-O octahedral when the two apical O atoms move away from the Mn ion [30]. As a result, the C-type antiferromagnetic order would occur in LCMO films if the compressive strain on ab-plane is strong enough. This point has been proved by many experiments [30]. As stated above, except the influence on average Mn-O-Mn bond distance and angles, the compressive strain can also compete with the $J-T$ distortion and lattice distortion caused by tolerance factor and thus magnetic structure of LCMO film. Therefore, the field induced strain effect in our experiments could affect the lattice distortion of LCMO films and thus contribute to the decreasing of magnetization. Theoretical and experimental studies have revealed that coexistence of multiple phases (e.g., ferromagnetic (FM), antiferromagnetic orbital order (AFM-OO)) and mutual interaction between them are very important in manganites as stated before [29, 31]. The electric field induced compressive strain and biaxial strain favor the formation of AFM-OO as well as changing the relative volume fraction of FM and AFM-OO region, and then the magnetization decreases. Although the strain in films is nonuniform due to the elastic strain in the film decreases with THE increasing of thickness, we can also use the strain value to denote the average strain state of films. The bulk and biaxial strain are determined by $\varepsilon_B = 1/3(\varepsilon_{xx} + \varepsilon_{yy} + \varepsilon_{zz})$ and $\varepsilon_{bi} = 1/4(2\varepsilon_{zz} - \varepsilon_{xx} - \varepsilon_{yy})$ respectively.11 In our case, the bulk and biaxial strain is about -0.026% and +0.051% respectively at E=9.6 kV/cm. That is to say the 0.051% biaxial strain will cause more than 8% decreasing of magnetization in LCMO/PMN-PT structures. Besides strain effect, the polarization effect induced by electric field should also be employed to discuss the field direction dependence of $\Delta M/M$. It is known that the charge accumulation or depletion has some impact on the magnetization in manganite. In LCMO compounds, the increasing of effective hole density near the interface will be beneficial to the enhancement of magnetization as the doping level small than 0.33. Moreover, the accumulation or depletion caused by polarization effect could also affect the relative volume fractions of FMM and AFM-OO phase in LCMO film as discussed afore. Hence, the magnetization of LCMO film with negative bias field will be higher and more robust than that with positive bias. Our results shown in Figure **2.10** are consistent with this point. The sharp, stable and reversal magnetization controlling effect caused by electric field observed here reveals that the electric field can tune the lattice distortion and effective carriers density artificially as well as magnetic structure by using manganite/PMN-PT structure. The change of magnetization due to the co-action of applied electric field induced strain and polarization effect observed here may be of practical interest. The magnetoelectric coupling coefficient $\alpha = \mu_0 \Delta M/\Delta E$ for the LCMO/PMN-PT structure is about 3.4×10^{-8} sm^{-1} which is similar to the results reported by C. Thiele et al [24] and smaller than that obtained in $La_{0.67}Sr_{0.33}MnO_3/BaTiO_3$ heterostructure [14]. Basing on the discussion above, one can deduce simply that the coupling between magnetic and lattice degrees of freedom is about -0.3285 emu (-0.1512 μ_B/Mn) and the coupling between magnetic and electric degrees of freedom is about 0.0255 emu (0.011 μ_B/Mn) in our LCMO films under the action of electric field 9.6kV/cm.

In results as shown above, one point worthy of special attention is that the electric field induced strain effect is dominant for the magnetization at low temperature, while the polarization is dominant for the electrical transport properties as shown in Figure **2.9**(b). This point implies that the magnetic structure of LCMO films is more sensitive to the local distortion of Mn-O octahedral than to the change of effective carriers density. Although the magnetization drops about 8%, the LCMO is still in ferromagnetic metal state and the decreasing of magnetization has little influence on the double exchange effect and percolative behavior as well as the electrical transport process. These features are consistent with previous results [15]. As mentioned in those works, the resistivity of manganites ρ varies with carrier density n and changing of magnetization ΔM in an empirical relation of $\rho \propto 1/nexp[-\Delta M/M_s]$, where Ms is the saturation magnetization. This relation reveals that the resistivity varies faster with the change of effective carrier density n than with relative change of magnetization ΔM as $\Delta M/M_s < 1$. In this work, we only give a possible explanation for the observed field control of the magnetic and electrical properties of LCMO/PMN-PT structures. To understand the co-action of strain and polarization effect comprehensively, further experiments about electron-lattice coupling of the manganite films are required.

3. Magneto-tunability of Photon-induced Effect in Manganite Based Heterojunctions

As mentioned above, the salient feature of manganite is the strong coupling between charge, spin, and orbital degrees of freedom, which is at the foundation of a rich set of coexisting phases with similar ground state energies. The energy scale of the dominant interactions that determines the electronic phases in correlated-electron systems is on the order of an electron volt. But the energy

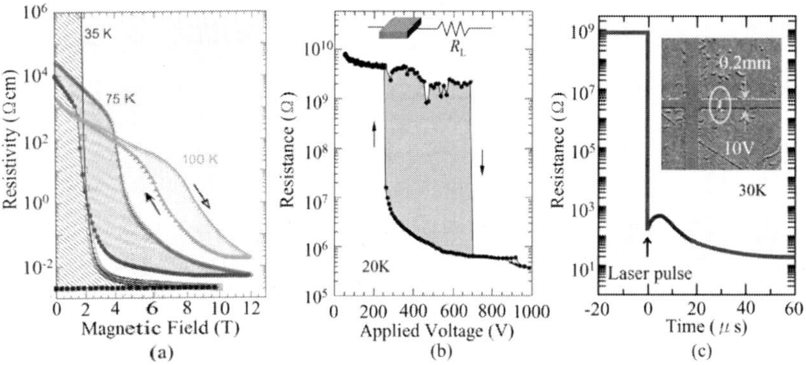

'igure 5.10. Metal-insulator phase control in the colossal-magnetoresistance ompound $Pr_{0.7}Ca_{0.3}MnO_3$. (a) A small increase in an external magnetic field vill lower the resistivity sharply, depending on temperature. The material also xhibits hysteresis in the resistivity (by increasing and then decreasing the mag-etic field). (b) A huge abrupt resistance change (and hysteresis) also accom-any voltage changes applied to a circuit connecting the sample in series with 1-MΩ resistor. (c) A 5-ns laser pulse creates a resistance change of almost orders of magnitude while a voltage is applied between gold electrodes. The hange in surface reflectance at the local metal-insulator transition induced by ne laser pulse shows up as a white streak (circled) between the 0.2-mm distance eparating the electrodes (see Inset picture) (Reproduced with permission from Reference [33]. Copyright 2003, AIP Publishing LLC).

36], $La_{0.7}Ca_{0.3-x}Ba_xMnO_{3-\delta}$ [37], and $Pr_{2/3}Sr_{1/3}MnO_{3-\delta}$ [38], in which the rigin of PPC was suggested to be associated with the oxygen deficiency of the amples. Moreover, the PPC effect induced by visible light has also been found n stoichiometric $(La_{0.3}Nd_{0.7})_{2/3}Ca_{1/3}MnO_3$ thin films.

Compared with conventional semiconductors, the manganites have some eatures such as the strongly correlated electronic state and the magnetic state ependent band structure. Hence, one may anticipate that the properties of the hotoconductivity in manganite films can be magnetically tunable, which is of pecial interest from the viewpoint of applications. Our group performed a com-rehensive study on the effect of the magnetic fields on the photoconductivity n $La_{2/3}Sr_{1/3}MnO_{3-\delta}$ (LSMO) thin films. The results show that the photocon-uctivity can be easily erased by the magnetic fields partially shown as follow.

difference that distinguishes each of the phases-metallic versus insulat ferromagnetic versus antiferromagnetic, is usually much smaller. Conseq mini-perturbations in an input signal can control a quickly switching ele phase output.

In CMR, a magnetic field aligns the spins on Mn sites, and that alig influences the conduction of electrons through adjacent oxygen atoms effect transforms the material from a charge-ordered/orbital-ordered (C insulator into a ferromagnetic metal, with a dramatic change in resisti addition, in what is termed as the magnetochromism effect, the magnet induced transition can also change the optical reflectivity spectrum of th rial on an energy scale up to 3 eV [32].

The most dramatic examples of such phase tuning occur in mate which the two competing electronic phases form a bicritical point. E magnetic fields are not the only way to control the metal-insulator tra near the bicritical point. Other triggers may do the trick: photoexcitation irradiation, electric fields, and impurity-atom doping, for example, can the electronic phase in an exotic way on ultrafast timescales. Tokura et a presented that the CO-OO to FM phase transformation in $Pr_{0.7}Ca_{0.3}Mn$ applying a magnetic field [33], an electric field [34], and a pulsed light shown in Fig. 5.10. The induced metal-insulator transition appears to local in the whole sample volume, but the change of resistance is still hu

3.1. Erasure of Photoconductivity by Magnetic Field in Man Films

Among the dynamic properties induced by external stimuli, the photoi effect, such as photoconductivity and photoinduced metal-to-insulator tion (MIT), is especially interesting from the viewpoint of photo-electri cations. There are two kinds of photoconductivity. One is transient ph ductivity (TPC), and the other is persistent photoconductivity (PPC). Fo the photoconductivity increment disappear quickly once the light illum is removed. For PPC, whereas, the photoinduced conductivity increme sists after the illumination, often with immeasurably long-time constants has been observed in many semiconductors. The PPC is usually interp terms of the separation of photogenerated electron-hole pairs to suppre recombination due to somewhat macroscopic, or microscopic barriers. T effect has been found in perovskite manganite thin films of $La_{2/3}Sr_{1/3}M$

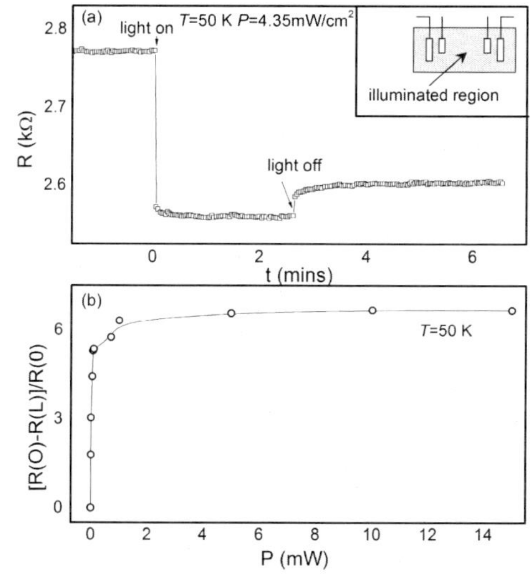

Figure 5.11. (a) Photoinduced resistance changes for oxygen-deficient LSMO films at temperature of 50 K. Inset shows experimental current and voltage contact configuration with respect to the illuminated region. (b) Power density dependence of PPC at temperature of 50 and 100 K respectively (Reproduced with permission from Reference [45]. Copyright 2006, AIP Publishing LLC).

In this study, the LSMO films were grown on Si (001) wafers by the off-axis DC-magnetron sputtering technique with the substrate temperature of 750 ^{o}C. For the photoconductivity effect measurement, the film samples were mounted in a vacuum room which was sealed by a double quartz glass window, and a He-Ne laser of the wavelength λ= 632.8 nm with variable power density was used to illuminate on LSMO film surface perpendicularly through the window. Simultaneously, the magnetic fields in the range of 0-1 T were supplied by a water-cooled magnet. A schematic illustration for the current and voltage contact configuration with respect to the illuminated region is shown in Fig. 5.11.

The typical photoconductivity results of oxygen-deficient LSMO film has been shown in Fig. 5.11. The time dependence of the resistance before, during and after laser illumination shows that the resistance decreases sharply and reaches a stable value with light illumination. Moreover, the photoinduced resistance drop does not return to the previous value but recovers partly upon stopping the laser illumination, which characterizes the PPC effect. This excited state can stay for hours if the sample is kept at low temperatures. These

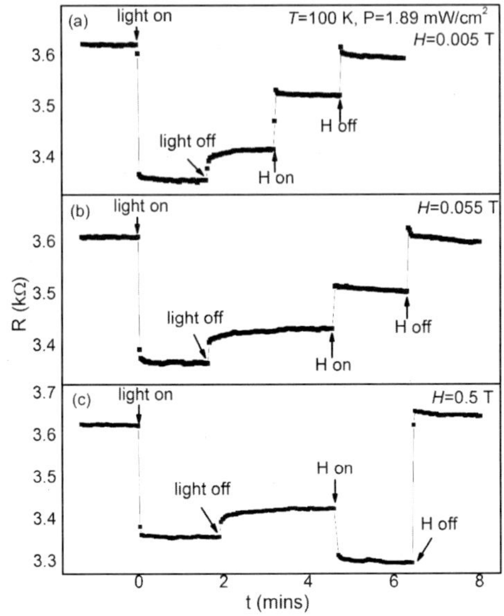

Figure 5.12. Time dependence of the resistance of oxygen-deficient LSMO films with the same light power density of 1.89 mW/cm^2 and different magnetic fields of (a) 0.005, (b) 0.055 and (c) 0.5 T (Reproduced with permission from Reference [45]. Copyright 2006, AIP Publishing LLC).

results reveal that both transient photoconductivity (TPC) and PPC exist in the LSMO films. In order to simplify the denotation of the resistances at different situations, the R_0, $R_L(on)$, and $R_L(off)$ were refered as the resistance before, during and after light illumination respectively. Therefore, it is easy to obtain that the magnitude of TPC [defined as (R_0-$R_L(on)$)/R_0] and PPC [defined as (R_0-$R_L(off)$)/R_0] are 7.6% and 6.08% respectively at 50 K with light density of 4.35mW/cm^2. Further experiments show that the PPC effect is only observed at low temperatures, where the sample lies in ferromagnetic (FM) metallic state. The PPC effect vanishes at around 200 K, which is close to Curie temperature (T_C ~200 K) of oxygen deficient LSMO films. It implies that there is intimate relation between PPC effect and magnetism state of LSMO films. The original state will be recovered when the sample is warmed back to the room temperature and then cooled again.

Figure 5.11(b) depicts laser power density dependence of PPC at temperature of 50 and 100 K. It shows that, for both 50 and 100 K, the PPC reaches

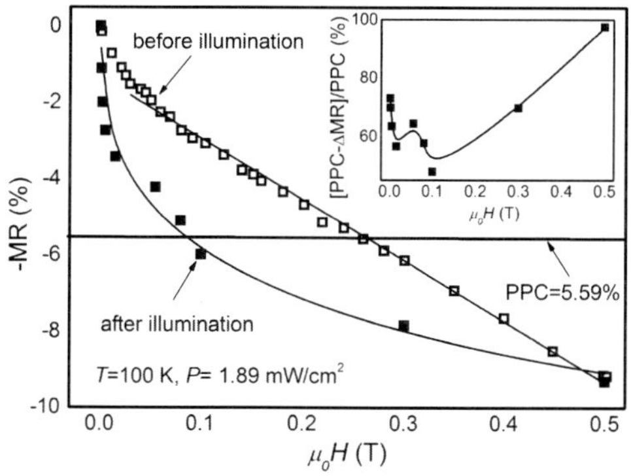

Figure 5.13. Magnetic field dependence of magnitudes of MR of LSMO films at temperature of 100 K. Open and filled squares are the MR data gained from resistance measurement before and after 1.89 mW/cm^2 laser illumination respectively. The inset shows [PPC(T)-ΔMR(T)] /PPC(T) as function of magnetic fields for light intensity of 1.89 mW/cm^2 at temperature of 100 K (Reproduced with permission from Reference [45]. Copyright 2006, AIP Publishing LLC).

saturation rapidly at the weak light density about 13 mW/cm^2, which is considered as a result of the maximum absorption of oxygen-deficient LSMO films. As for appreciably decreasing of PPC above 21.8 mW/cm^2 at 100 K, it is contributed to little heating effect because resistance increases fast with heating at the temperatures close to the T_C.

To test whether the PPC observed in the oxygen-deficient LSMO films can be influenced by applied magnetic fields, a magnetic field parallel to the films plane was utilized. The results are shown in Fig. 5.12. It displays that the resistance of films stays at a stable value after a constant light illumination. However, as a small magnetic field of 0.005 T and 0.055 T parallel to the film surface is applied, the resistance shows a quick increase as shown in Fig. 5.12(a) and (b). When the magnetic field of 0.5 T is applied, the resistance of films shows a sharp decrease contrarily as shown in Fig. 5.12(c). Moreover, resistance can return to the original value if the external field is removed, i.e., both PPC and magnetoresistance (MR) disappear. These results reveal that the PPC can be erased mostly by the magnetic field easily and make the samples back to original state, which is interesting from the viewpoint of applications in devices. Similar phenomenon has been observed in the whole temperature range in which PPC effect exists.

To further explore the erasure of PPC by the magnetic field, the value of MR (defined as $(R_0\text{-}R_H)/R_0$) before and after light illumination is investigated together carefully. Fig. 5.13 shows MR as a function of magnetic fields at 100 K with very weak light density of 1.89 mW/cm^2. The PPC is about 5.59% at 100 K with light density of 1.89 mW/cm^2 as shown by transverse line in Fig. 5.13. Open and filled squares in Fig. 5.13 present the MR of LSMO films before and after the illumination of laser respectively. Before the irradiation, a fast rise in MR ratio is observed in the low-field region of H <30 mT, whereas the variation of the MR is nearly linear with increasing of the field and can be described by:

$$MR = \beta + \gamma H \tag{3}$$

where β and γ are coefficients and the $\gamma \sim 0.16$/T. Furthermore, the MR ratio about 9% is achieved in the field of 0.5 T. After the illumination, the behavior of the MR is quite different and is nonlinear in the whole measurement range. The different relation between MR and the magnetic field before and after light illumination indicates the existence of different electrical transport properties of LSMO films. Another point worth of special attention is that the magnitude of MR ratio of LSMO films after laser illumination is larger than that before illumination within whole the measurement range. It reveals that the PPC effect cannot be erased completely by the magnetic field.

In order to present the proportion of PPC erased by the magnetic field, we define the expression $[PPC(T)\text{-}\Delta MR(T)]/PPC(T)$, in which $\Delta MR(T)$ is the absolute value of the difference between MR after illumination and MR before illumination at certain temperature. The inset of Fig. 5.13 shows the proportion of PPC erased by the magnetic field as a function of magnetic fields. It shows that the proportion of erased PPC is 73% in magnetic field of 0.001 T and 97.5% in 0.5 T. The nonlinear behavior of the proportion reveals the complex electrical transport in LSMO films after applying the magnetic field.

3.2. Effect of Bias Current on the Resisitivity and Photoconductivity in Manganite Films

As mentioned above, the electric field, except magnetic field and light illumination, can also shift the delicate balance between different phases in strong correlated manganites and may induce dramatic changes in their physical properties. The electric field induced sudden drop of resistivity in $Pr_{0.7}Ca_{0.3}MnO_3$ shown in Fig. 5.10 is one of the examples. Subsequently, many experiments performed on charge-ordered samples $Pr_{1-x}Ca_xMnO_3$ involving current injection into highly conducting filamentary paths. Moreover, several works presented that such electric field or current effect on manganites not only can be observed in charge-ordered system but also in compounds with ferromagnetic ground state, such as $La_{0.8}Ca_{0.2}MnO_3$ [39], $La_{0.85}Ba_{0.15}MnO_3$ [40], and $La_{0.67}Ca_{0.33}MnO_3$ [41] thin films.

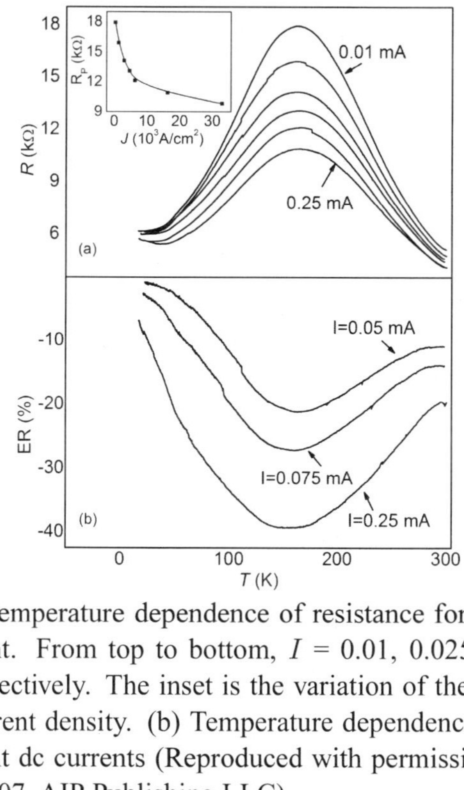

Figure 5.14. (a) Temperature dependence of resistance for LSMO film under different dc current. From top to bottom, I = 0.01, 0.025, 0.05, 0.075, 0.1, and 0.25 mA, respectively. The inset is the variation of the peak resistance as function of dc current density. (b) Temperature dependence of ER for LSMO film under different dc currents (Reproduced with permission from Reference [44]. Copyright 2007, AIP Publishing LLC).

Both the electric field and light illumination can produce the change of resistivity of manganite films [43, 42]. However, the relationship between the photoconductivity and applied electric current (field) in the manganese oxides, which is very important in realizing the useful application of ER and photoconductivity (PR) in these compounds in the future, is not very clear until now. To understand this point partially, our group performed studies on the effect of the electric current on the resisitivity and photoconductivity in an oxygen-deficient $La_{2/3}Sr_{1/3}MnO_{3-\delta}$ (LSMO) thin films [44]. The photoconductivity of LSMO films and its magnetic field effect has been studied before(Fig. 5.11 and Fig. 5.12) [45]. Before we study the relationship between electric field-induced effect and photoinduced effect in manganite, the electric field effect on LSMO films was studied separately at first.

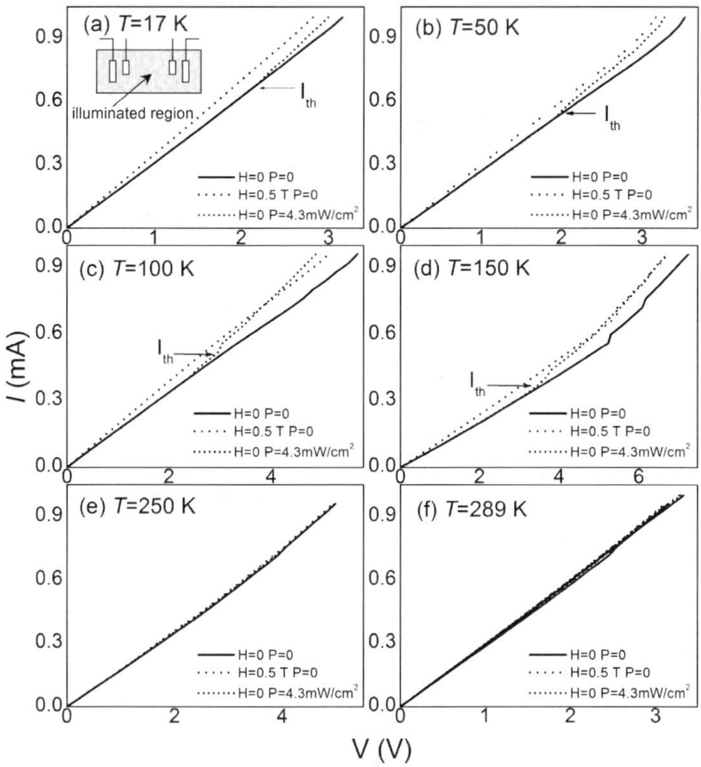

Figure 5.15. $I-V$ characteristics at different temperatures of oxygen-deficient LSMO films by tuning bias currents. Inset shows experimental current and voltage contact configuration with respect to the illuminated region (Reproduced with permission from Reference [44]. Copyright 2007, AIP Publishing LLC).

Fig. 5.14(a) shows the temperature dependence of resistance of LSMO under different dc currents. There exists a giant drop of resistance of LSMO film with the increasing of applied electric current. Moreover, the temperature where

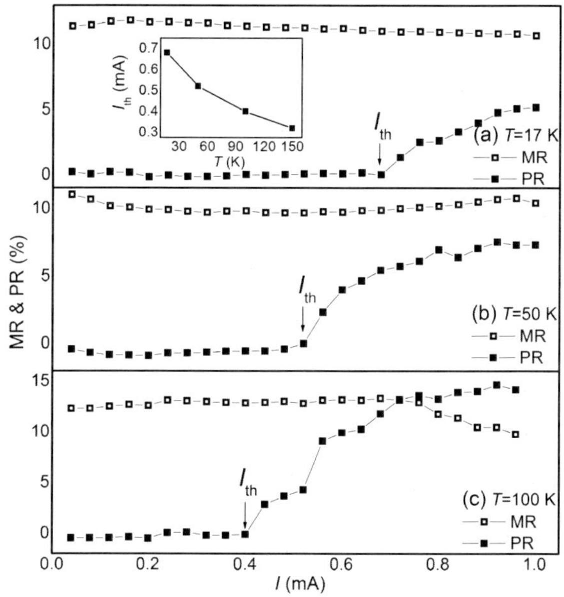

Figure 5.16. MR (defined as $[R(0)$-$R(H)]/R(0))$ for P =0, H =0.5 T and photoinduced resistance (PR) (defined as $[R(0)$-$R(L)]/R(0))$ for P =4.3 mW/cm^2, H =0 as functions of applied current at temperature of (a) 17, (b) 50 and (c) 100 K. The inset shows the temperature dependence of I_{th} (Reproduced with permission from Reference [44]. Copyright 2007, AIP Publishing LLC).

occurs metal-insulator transition T_P do not move within whole measurement current range. Fig. 5.14(b) describes the temperature dependence of ER under different currents. The ER, defined as $[R(I)$-$R(0.01mA)]/R(0.01mA)$, is negative because the current reduces the resistance. It is also noted that ER is maximum at T_P and reaches -39.3% with T=160 K and I = 0.25 mA. ER becomes very small at low temperatures. The inset of Fig. 5.14(a) is the variation of the peak resistance R_P with dc current density. It shows a nonlinear behavior and can be fitted exponentially by R_P=8736.4exp(-J/4.28)+10239.8, the unit of J is 103 A/cm^2. Such an exponential decreasing of R_P with current is consistent with the results in ref [41]. It also should be pointed out that the value of the resistance comes back up sharply again when the current is reduced, indicating the effect is reversible.

A difficult problem which one unavoidably faces upon in the process of investigating the effect of electric current/field on resistance is Joule heating.

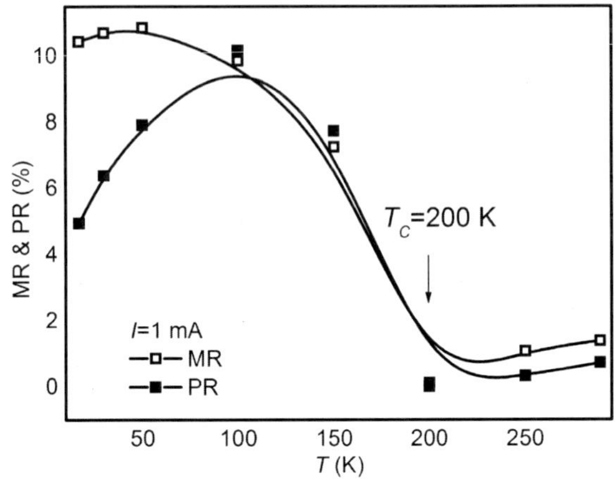

Figure 5.17. MR under a field of 0.5 T and PR at a light density of 4.3 mW/cm^2 at I = 1 mA as functions of temperature (Reproduced with permission from Reference [44]. Copyright 2007, AIP Publishing LLC).

If the heating effect is severe because of poor heat diffusion, T_P of LSMO film would shifts to lower temperatures as mentioned in ref [40], in which the heating problem hinders the visibility of the intrinsic current effect. As for our results, Fig. 5.14(a) clearly shows that T_P remains unchanged with increasing the electric current and the peak resistance drops remarkably. The insensitivity of T_P to current differs from the dependence of T_P on magnetic field as the magnetic field shifts T_P to higher temperatures. The value of ER is comparable to magnetoresistance (MR) induced by a strong magnetic field of several Tesla.

As observed in Fig. 5.11 and Fig. 5.14, magnetic field, electric field/current, and light illumination can all induce the change of resistivity of LSMO films. Then, let's shift the interaction between these effects. The current versus voltage ($I - V$) with and without light illumination and magnetic field respectively of the oxygen-deficient LSMO films are measured by tuning the applied current, and the typical results are presented in Fig. 5.15. At high temperature (above T_C), as seen in Fig. 5.15(e) and (f), all $I - V$ curves in different conditions are close to each other, which indicates the applied magnetic field and the light illumination hardly affects electrical transport behavior of LSMO films above T_C. As $T < T_C$, the transport of LSMO film behaves is a different way. As seen from Fig. 5.15(a), (b), (c) and (d), the $I - V$ curves under H =0 and 0.5

T are nearly linear. However, the voltage for H =0.5 T is smaller than that for H=0 T at the same current bias, i.e., behaving as a negative MR which is consistent with results in Fig. 5.14. An intriguing phenomenon occurs when the laser illumination is applied. It is found that the $I-V$ curves for H =0 and P =4.3 mW/cm^2 are as same as curves for H =0 and P =0 under low bias currents. However, as the current bias exceeds a threshold I_{th}, the $I-V$ curves for H=0 and P=4.3 mW/cm^2 deviate from original orientation and approach the curves for H=0.5 T and P=0 gradually. That is to say, the photoconductivity effect only appears as $I > I_{th}$ and $T < T_C$. These results imply that the photoconductivity effect in oxygen-deficient LSMO films has an intimate relation not only with magnetic state of the sample itself but also with the magnitude of the applied bias current. Further calculation indicates that the MR (defined as for P = 0, H = 0.5 T) ratio almost keeps a constant within the whole current range at a fixed temperature. However, the photoinduced resistance ratio (PR) (defined as for P=4.3 mW/cm^2, H=0) increases sharply from zero as $I > I_{th}$. This point can be clearly reflected as in Fig. 5.16 (a, b, c), in which both the MR (defined as for P = 0, H = 0.5 T) and the photoinduced resistance (PR) (defined as for P=1 mW, H=0) as functions of applied current are presented at T = 17, 50 and 100 K respectively. It shows clearly that the PR ratio increases sharply from zero as $I > I_{th}$ and increases with increasing of bias current. This switch effect may be of worth in the potential applications of new generation of data storage and sensing devices. The inset of Fig. 5.16 shows the temperature dependence of I_{th}. It is found that the I_{th} of current-assisted photoconductivity effect increases with decreasing of temperature. It can be understood by considering the insulator phase in LSMO film becomes more stable with the decreasing of temperature. Therefore, it needs larger applied current density to change it into metastable phase which can be affected by light illumination. Another point worthy of noting in Fig. 5.15 is that there are some kinks in the I-V curves at the temperature of 150 K as shown in Fig. 5.15(d). It is considered as a result of emergence of prominent ER near the T_P with increasing of external electric current which is consistent with results of resistance measurement under different electric current as shown in Fig. 5.14.

In order to explore the variation of MR and PR at different temperatures further, MR under the magnetic field of 0.5 T and PR under the light density of 4.3 mW/cm^2 at a selected bias current of 1 mA have been studied, and the results are shown in Fig. 5.17. It is found that the maximum of MR is 10.8% at T = 50 K and H = 0.5 T, and that the PR achieves 10.2% at T = 100 K with the

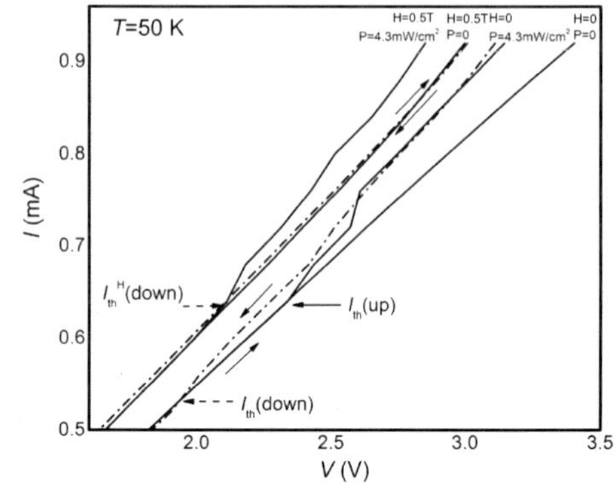

Figure 5.18. $I - V$ characteristics of the LSMO films at T= 50 K (Reproduced with permission from Reference [44]. Copyright 2007, AIP Publishing LLC).

light density of 4.3 mW/cm^2. One point deserving special attention is that both MR and PR rise sharply as T is lower than T_C of LSMO films ($T_C \sim$ 200 K). These results reveal that there exists a close relation between electrical transport and magnetization in oxygen-deficient LSMO films.

As stated above, the $I - V$ curves of LSMO film show discontinuous behavior and the resistance of LSMO film changes from higher resistance state to lower resistance state under the co-action of light illumination and bias current at low temperatures. To further explore this current-assisted photoconductivity effect of LSMO film, the $I - V$ measurements at different conditions were carried out. The typical results at temperature of 50 K have been shown in Fig. 5.18. There are two points worthy of special attention in Fig. 5.18. The first one is that there exists hysteresis in the $I - V$ curves with light illumination. Both the branch curve of sweeping current up and down exist threshold current I_{th} where photoconductivity effect emerges. It is found that the I_{th}(down), when current sweeps down, is smaller than I_{th}(up). That is to say the higher threshold current is necessary for LSMO films to change from higher resistance state to lower resistance state with sweeping current up. As for the current sweeping down, the lower resistance state can keep longer until the current is lower than I_{th}(down). Otherwise, the hysteresis behavior of LSMO films is negligible when the magnetic field is applied alone as shown in Fig. 5.18. The second one is that only MR exists as $I < I_{th}^H(down)$, and photoconductivity present itself as $I > I_{th}^H(down)$ with H = 0.5 T and P=4.3 mW/cm^2. The accumulation of MR and photoconductivity proves that there exist difference between MR and photoconductivity. It is obviously that the I_{th}(down) increases from 0.52 to 0.64 mA as a magnetic field 0.5 T is applied and the magnitude of resistance reduction MPR (defined as $[R(0)\text{-}R(H, L)]/R(0))$ with H = 0.5 T and P=4.3 mW/cm^2 is smaller than the summation of PR (H = 0 T, P =4.3 mW/cm^2) and MR (H = 0.5 T, P=0 mW/cm^2). For example, the MPR is 14.6% at current bias of 1 mA, which is smaller than the summation of PR ($\sim$ 7.31%) and MR ($\sim$ 10.42%). These results reveal that the effects on conductivity by the magnetic field and the light illumination respectively are not independent completely but correlative with each other.

3.3. Magnetic Field Effect on Photovoltage in Manganite-based Heterojunction

As mentioned in Chapter **3**, lots of attempts were made to fabricate heterojunctions using manganites and other perovskite oxides. It has been found that fabricated heterojunctions can have excellent rectifying property through properly controlling dopant species and the doping level of the oxides [9]. On the other hand, compared with conventional semiconductors, the manganites have some features such as a strongly correlated electronic state and a magnetic state dependent band structure. Hence, one may anticipate that the properties of manganite-based heterojunctions can be tuned mangnetically, which is of special interest from the applications viewpoint. In 2004, Sun et al. reported that manganite-based heterojunctions were magnetically tunable, i.e., the diffusion/breakdown voltage and the junction resistance underwent a great change under the action of external magnetic fields [10]. At the same time, Sheng et al. start to explore the magnetic field effect on photovoltaic effect (PVE) of the manganite-based heterojunctions [46, 47].

As is well known that the PVE of the conventional *p-n* junction is widely used in photocells, photosensors etc. This PVE is associated with the generation and diffusion of extra carriers generated by exposure to light. That is to say, the *p-n* junction when illuminated can generate extra holes and electrons (called photocarriers) if the photon energy is larger than that of the semiconductor band gap. The drift of photocarriers towards p and n sides under the built-in electric field in the depletion layer of the *p-n* junction can produce a PVE. As discussed in Chapter **4**, the PVE has been discovered in a manganite-based heterojunction. Although the PVE mechanism is not very clear at present, it is apparent that the

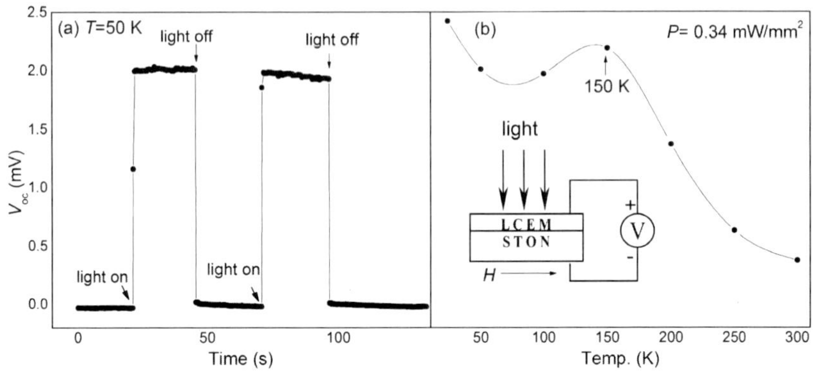

Figure 5.19. (a) Response of photovoltage to light illumination measured at a constant temperature of 50 K under the light power density of 0.34 mW/mm^2 (λ=632.8 nm). Inset is a schematic illustration of the PVE measurement under magnetic fields. (b) Temperature dependence of photovoltage V_{OC} under the light power density of 0.34 mW/mm^2 (Reproduced with permission from Reference [46]. Copyright 2005, AIP Publishing LLC).

PVE is related to the magnetic behaviors of manganite films in the heterojunction. We know that the magnetic behavior of manganite films can be changed by external magnetic fields. So, it should be of significance to test whether the observed PVE can be manipulated by the external magnetic field. At the first stage, Sheng et al. performed a detail study on the effect of external magnetic field on the PVE in a $La_{0.7}Ce_{0.3}MnO_3/SrTiO_3$-Nb (LCEM/STON) heterojunction.

The response of open-circuit photovoltage (V_{OC}) to light illumination measured at temperature T=50 K (light power density P=0.34 mW/mm^2) was presented in Fig. 5.19(a). It shows that the LCEM/STON junction exhibits an obvious PVE, and the V_{OC} shows a quick switch between $\sim$ 0 and $\sim$2.0 mV corresponding to "light on" and "light off" similar to the results previously reported [48]. The temperature dependence of the V_{OC} for a fixed laser power density of P = 0.34 mW/mm^2 is shown in Fig. 5.19(b). It reveals a nonlinear increase of V_{OC} as the temperature decreases, and a local maximum appearing near 150 K.

To test whether the observed PVE in the LCEM/STON heterojunction can be influenced by applied magnetic fields, an external magnetic filed of 0.5 T was applied on the heterojunction. The results are shown in Fig. 5.20. Without

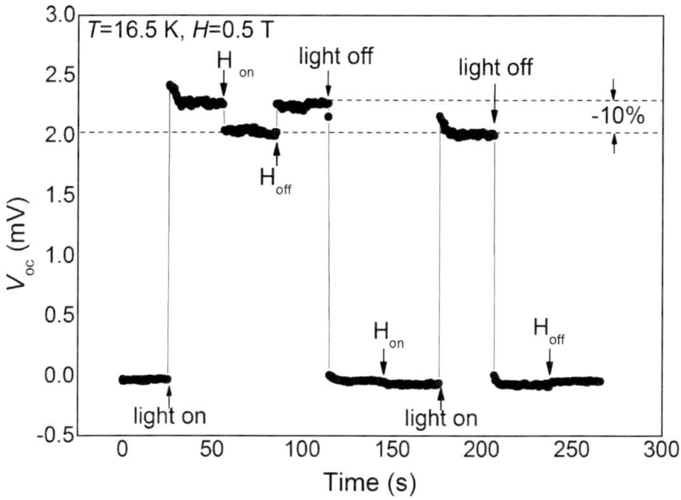

Figure 5.20. Response of photovoltage to light illumination measured at a constant temperature of 16.5 K with the light power density of 0.34 mW/mm^2 and at an applied magnetic field of 0.5 T parallel to the film surface of LCEM/STON (Reproduced with permission from Reference [46]. Copyright 2005, AIP Publishing LLC).

magnetic fields, V_{OC} stays at a stable value under the constant light illumination. However, as an external magnetic fields of 0.5 T parallel to the film surface is applied, V_{OC} shows a quick decrease, and its relative variation [defined as $[V_{OC}(0)\text{-}V_{OC}(\text{H})]/V_{OC}(0)\times 100\%$] is larger than 10% at T = 16.5 K. Moreover, V_{OC} can return to the previous value if the external field is removed, that is, the dependence of the V_{OC} on the magnetic fields is reversible. Similar phenomenon has been observed in the whole measured temperature range. In order to investigate the effect of the external magnetic fields on the PVE further, we apply an external magnetic fields to the film before light illumination. Figure 5.20 shows that the magnitude of V_{OC} also keeps $\sim$10% below the V_{OC} value without external filed. These results reveal unambiguously that the external magnetic fields can manipulate the PVE of a LCEM/STON heterojunction. Similar phenomenon is also observed in other manganite-based heterojunctions such as $La_{0.32}Pr_{0.35}Ca_{0.33}MnO_3$/STON, which indicates the magnetically tunable PVE is a common feature of the manganite-based heterojunctions.

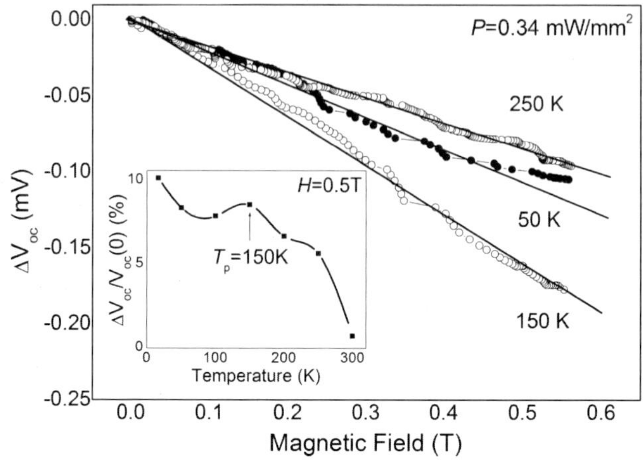

Figure 5.21. Variation of photovoltage ΔV_{OC} as function of magnetic field at selected temperatures 50, 150 and 250 K. The inset plot shows the temperature dependence of relative variation of photovoltage $\Delta V_{OC}/V_{OC}(0)\times 100\%$. The used light power density is 0.34 mW/mm^2 (Reproduced with permission from Reference [46]. Copyright 2005, AIP Publishing LLC).

The external magnetic field dependence of V_{OC} for constant temperature and light power density has also been explored. Figure 5.21 shows the variation of photovoltage (ΔV_{OC}) with magnetic fields for selected temperatures 50 K, 150 K, and 250 K. ΔV_{OC} decreases monotonously with applied fields, and the field dependence of ΔV_{OC} can be described as following equation:

$$V_{OC}(H) = V_{OC}(0) - \alpha(T)H \tag{4}$$

where V_{OC}(H) and V_{OC}(0) represent the photovoltage with and without magnetic fields, respectively, H is magnetic fields intensity and $\alpha(T)$ is a temperature dependent parameter. For a constant magnetic field of 0.5 T, the temperature dependence of $\Delta V_{OC}/V_{OC}(0)$ is plotted in the inset of Fig. 5.21 in the temperature range of 20-300 K. It shows that $\Delta V_{OC}/V_{OC}(0)$ does not change monotonically with temperature and has a bump near 150 K. It is interesting that $\Delta V_{OC}/V_{OC}(0)$ and V_{OC}(T) curves show a similar temperature-dependent feature (Figs. 5.19(b) and 5.21). In other words, both of them have a bump around 150 K. The magnetization of LCEM/STON heterojunction under exter-

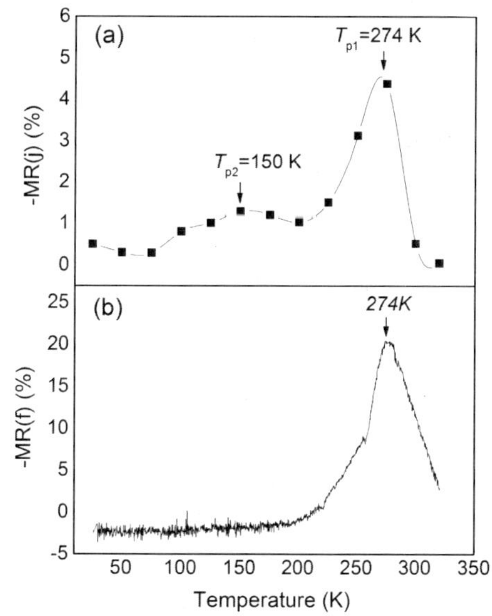

Figure 5.22. Temperature-dependent magnetoresistance of (a) LCEM/STON heterojunction $[R_j(0)\text{-}R_j(\mathrm{H})]/R_j(0)\times 100\%$ and (b) LCEM film $[R_f(0)\text{-}R_f(\mathrm{H})]/R_f(0)\times 100\%$ (Reproduced with permission from Reference [46]. Copyright 2005, AIP Publishing LLC).

nal magnetic fields of 0.001 T and 0.5 T was measured in the temperature range of 5-300 K. However, it is difficult to find some unambiguous relation between ΔV_{OC} and ΔM as stated in [46].

Without detail information of interfacial information, such as structure and electronic state, it was difficult to obtain the exact origin of the magnetic dependence of V_{OC} observed in LCEM/STON heterojunction. However, one may have chance to deduce the possible influencing factors from the correlation phenomena. As is known in conventional junction, the V_{OC} has a close relation with the built-in potential (V_{bi}) at the interface. Therefore, the field effect on V_{OC} may be a result of the variation of the V_{bi} in the depletion layer under magnetic fields. In order to test whether this assumption is true, Sheng et al. performed the measurement of the resistance of the heterojunction (R_j) since built-in field can be reflected by R_j. The zero-bias junction resistance at zero magnetic field $R_j(0)$ and at an external magnetic field $R_j(\mathrm{H})$ were obtained through measuring their current-voltage ($I - V$) curves and calculating $I - V$ curves according to

$R_j = \partial V/\partial I$ for I=0. The temperature dependence of magnetoresistance (MR) of the heterojunction, defined as $MR(j)=[R_j(H)-R_j(0)]/R_j(0)$, is shown in Fig. 5.22(a). MR exhibits a negative signal in the whole measured temperature range with two local maximums located at T_{p1} = 274 K and T_{p2} = 150 K, respectively. Considering that the structure of the heterojunction is composed of LCEM film and the depletion layer and the substrate, the resistance of the heterojunction can be expressed as $R_j=R_f+R_d+R_s$, where R_j, R_f, R_d and R_s are the resistance of the heterojunction, the LCEM film, the depletion layer and the substrate, respectively. Noting that the behavior of STON is magnetic field independent, we attribute the MR of the heterojunction to the variation of the resistances of the LCEM film and the depletion layer under the magnetic field. In order to separate the MR contribution of the LCEM film and the depletion layer to MR(j) of the heterojunction, the MR(f) of the LCEM film, defined as $MR(f) = [R_f(H) - R_f(0)]/R_f(0)$, is measured at H = 0.5 T. The MR(f) as a function of the temperature is shown in Fig. 5.22(b). It shows that MR(f) has a maximum at T = 274 K, which is the same temperature as T_{p1} as shown in Fig. 5.22(a). In an approximate method, we calculated R_f ($\sim$0.5 Ω) and R_s ($\sim$0.02 Ω) from their resistivities of LCEM film and STON substrate.8 Therefore, the resistance of depletion layer R_d is about 90% of resistance of junction R_j and it is accordant with traditional *p-n* junctions. Therefore, combining Fig. 5.22(a) and (b), we attribute the $MR(j)$ in the vicinity of T_{p1} to the contribution both from the $MR(f)$ of the LCEM film (about$\sim$2%) and MR(d) of depletion layer(about$\sim$2.5%). On the other hand, the $MR(j)$ in the vicinity of T_{p2} should be ascribed to the contribution of the depletion layer only. As a result, the variation of the depletion layer under magnetic fields is suggested to be the underlying reason for the appearance of ΔV_{OC}.

From the discussion above, one point can be concluded that the variation of the properties of the depletion layer under the applied magnetic fields relates to the field effect on PVE observed in LCEM/STON. As for the field influence on the depletion layer, three factors have been considered [46]. The first factor is from LCEM film itself because its magnetic state can be changed by the applied magnetic fields. As pointed in Chapter **1**, the manganites undergo a change of band structure accompanying paramagnetic (PM) to ferromagnetic (FM) transition, i.e., the spin-up and spin-down sub-bands split in FM state of LCEM. Under the action of external magnetic fields, the spin disorder around curie temperature (T_C) is suppressed and both T_C and the insulator-metal transition temperature (T_{MI}) are shifted to higher temperatures, leading to the CMR effect. As a result, the resistance of LCEM layer is reduced in the vicinity of T_C under applied magnetic fields. The spin-polarized carriers tend to occupy the lower sub-bands, which results in the reduction of the Fermi level of LCEM. Therefore, the potential barrier between LCEM and STON is decreased. In other words, the depletion layer of LCEM/STON junction under external magnetic fields becomes effectively thinner, which results in the decrease of V_{OC}. This factor explains the appearance of ΔV_{OC} near the T_C because the reduction of Fermi level of LCEM and the potential barrier between LCEM and STON should only occur in the vicinity of T_C. However, the inset in Fig. 5.21 shows that the change in V_{OC} caused by the external magnetic field takes place in a wide temperature from 20 to 300 K. Hence, other factors should be considered. One should be considered is that there exist some FM insulating phases except for FM metallic phases in the depletion layer caused by the interface strain, i.e., there is phase separation in the interfacial region of the heterojunction as suggested by recent theoretical and experimental studies of manganites. The fractions of both FM metallic and FM insulating phases can be changed by applied magnetic fields, which results in the variation of the depletion layer. Another one may be the change of canting angle of magnetic domains under magnetic fields. Based on our experimental results, it appears to be more reasonable to attribute the V_{OC} change induced by the applied magnetic field to the later two factors. To prove this point, a further study on the manipulation of depletion layer is highly expected.

3.4. Magnetic Field Effect on Photocurrent in Manganite-based Heterojunction

With the development of the film-fabrication facilities and characterization technologies, new concepts and ideas on the manganite-based heterojunctions have been emerged. One of them is the "Correlated Solar Cell". According to this concept, the multiple carrier generation caused by the light illumination in correlated electron materials is utilized. In detail, as shown in Fig. 5.23, when a light is illuminated on a charge-ordered insulating state typically observed in transition-metal oxides, it is known that the collective activation of localized charges occurs leading to phase transition from an insulating into a metallic state [Fig. 5.23(a)]. During this photoinduced phase transition, many carriers are released from localized state to a itinerant state[Fig. 5.23(a)]. As is known that, in general, the correlated electron materials have large carrier den-

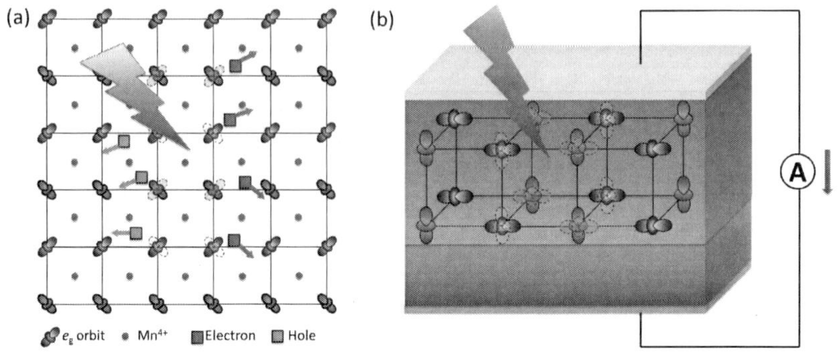

Figure 5.23. (a) A light is illuminated on correlated material with charge-ordered insulating state. A phase transition from an insulating into a metallic state is induced by the light illumination due to the collective activation of localized charges. And then multiple charge carriers are generated during this photoinduced phase transition due to the strong correlation.(b) A schematic image of the correlated solar cells.

sity ($\sim 10^{23}/cm^3$). Therefore, the multiple charge carriers are generated even with one photon excitation due to the light induced phase transition. It may be treated as **correlated multiple carrier generation**, which is a consequence of strong electron correlation.

With this in mind, one may consider that once this multiple carrier generation is applied to a solar cell, a higher quantum efficiency can be obtained. As known in conventional semiconductor solar cells, part of the energy supplied by the photons is lost as heat instead of being converted into electricity. Similar to the multiple carrier generation developed in semiconductor quantum dots, such correlated multiple carrier generation will be one of the ways to capture waste heat. Consequently, great improvement in photo-electron energy transfer can be expected[Fig. 5.23(b)]. Moreover, the correlated electron materials are sensitive to the external stimuli and the correlated solar cell based on these compounds might be external field tunable, which provides another freedom for the future applications. Absolutely, this concept on "correlated electron solar cell" has been attracting attentions as a potential next-generation solar cell.

The realization of such concept relies on producing an interface between surfaces with different electronic properties, called a heterojunction, to separate the delocalized/excited carriers. In Chapter **4**, it has been shown that when manganites come into contact with each other, photovoltaic phenomena can emerge due to competition between non-interacting, semiconducting electrons and a phase in which the electrons work togetherłknown as a correlated system. To realize the "correlated solar cell", our group paid a lot attentions on the manganite-based heterojunctions and their external field manipulation. In this program, the first compound our group chose was the $Pr_{0.55}(Ca_{1-y}Sr_y)_{0.45}MnO_3$ (PCSMO) with y=0.25. For bulk with such doping level, the PCSMO has a bicritical feature, i.e., it between the ferromagnetic-metallic (FMM) and the charge/orbital ordered states (CO/OOI), of the electronic phase diagram [49]. Near this bicritical point, the PCSMO crystal shows coexistence and competition of the two phases, which is sensitive to external stimuli [49]. In 2008, Matsubara et al. studied the photoinduced switching in PCSMO single crystal by irradiation of a femtosecond laser pulse [50]. They found that, for a low excitation density, the charge and orbital order is partially melted and the microscopic quasi-FM domains are generated, which rapidly return to the CO/OOI state. When the excitation density exceeds a threshold, the CO/OOI phase transforms permanently to the FM one. Moreover, they observed that one excitation photon changed 4-5 Mn sites from the CO/OOI state to the non-ordered or quasi-FM state. That is to say multiple carrier generation (4-5 carriers/photon) can happen in such PCSMO with bicritical feature. Furthermore, some theoretic works predicted that the multiple carrier generation by a single photon can be realized in correlated electron solar cells [51, 52, 53]. In these reports, correlated physics and conventional modeling based on semiconductor physics are well integrated. The minority carriers, i.e., electrons and holes, produced through this process can be separated by the built-in potential at the heterointerface and then collected as photocurrent [53]. These results are very exciting and inspire us to pursue the correlated solar cell with a high quantum efficiency.

As stated in Chapter **1**, one of the key features related to the phase transition in manganites is the phase separation between different electronic/magnetic states. This sort of phase separation is a common characteristic of many correlated electron oxides, originating from the strong coupling between lattice, charge, orbital, and spin degrees of freedom. The phase transition caused by external stimuli, to some extent, is the consequence of the phase competition and the re-balance of the ground state. From this viewpoint, it is necessary to clarify the existence and dynamics of phase separation in manganite films for the purpose of correlated solar cells.

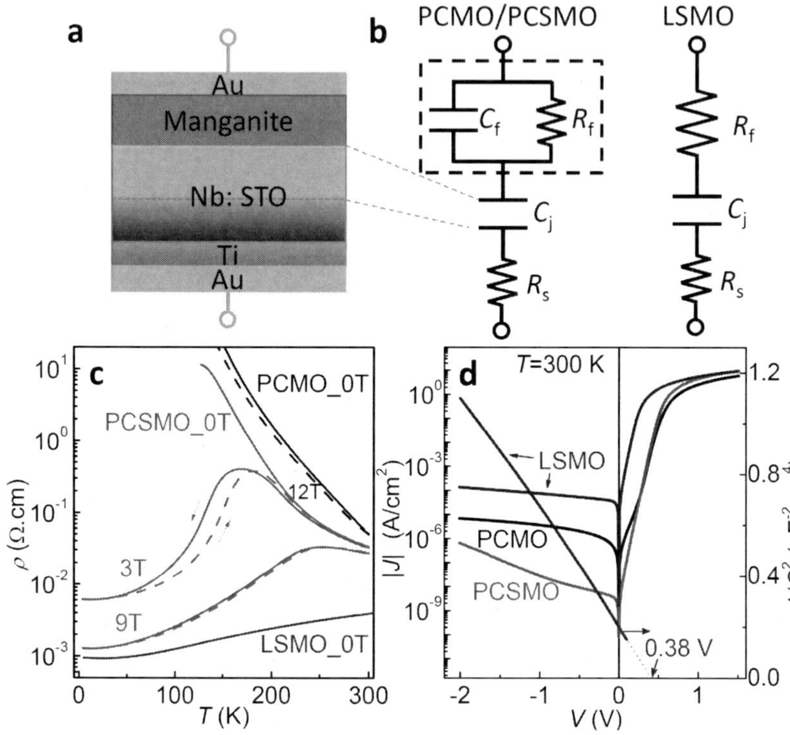

Figure 5.24. (**a**) junction configuration and (**b**) equivalent circuits for PCMO(PCSMO)/STON and LSMO/STON junctions. (**c**) Temperature dependence of in-plane resistivities for PCMO, PCSMO, and LSMO films under various external magnetic fields. The arrows indicate the direction of temperature variation. The dashed line was used to present the $\rho - T$ of PCMO film measured at 12 T field for clarity. (**d**) Current density-voltage (left axis) relations for three junctions and capacitance-voltage (right axis) characteristics for LSMO junction measured at 300 K. The dashed line is the extrapolation of liner fitting for $1/C^2 - V$ curve of LSMO junction with the intercept on voltage axis $\sim$0.38 V. In both **c** and **d**, the black, red, and blue curves present the data for PCMO, PCSMO, and LSMO samples, respectively. Here, PCMO, PCSMO, LSMO, and STON stand for $\mathrm{Pr_{0.5}Ca_{0.5}MnO_3}$, $\mathrm{Pr_{0.55}(Ca_{0.75}Sr_{0.25})_{0.45}MnO_3}$, $\mathrm{La_{0.65}Sr_{0.35}MnO_3}$, and Nb-doped (0.01 wt%) $\mathrm{SrTiO_3}$ (Reprinted by permission from Macmillan Publishers Ltd: Nature Communications (Ref. [28]), Copyright (2012)).

To reveal such property and its dynamics in the PCSMO films, our group designed a unique *p-n* junction configuration and performed a comprehensive dielectric dynamics measurements. In this case, $Pr_{0.55}(Ca_{0.75}Sr_{0.25})_{0.45}MnO_3$ (PCSMO) was studied. In addition to this sample, the half-doped $Pr_{0.5}Ca_{0.5}MnO_3$ (PCMO) with very rigid charge ordered (CO) state and the metallic $La_{0.65}Sr_{0.35}MnO_3$ (LSMO) have also been examined as the references. With epitaxially grown manganite films (~40 nm in thickness) on n-type Nb-doped (0.01 wt%) $SrTiO_3$ substrates (STON), *p-n* junctions were built for PCMO and PCSMO samples, while a Schottky junction was built for metallic LSMO sample, as shown in Fig. 5.24**a** and **b**. The temperature dependence of in-plane dc transport properties of the three manganites films measured in selected external magnetic fields are presented in Fig. 5.24**c**. Without magnetic field, the PCMO and PCSMO films are insulating while LSMO is metallic in the whole measurement temperature range. The CO transition temperature T_{CO} for PCMO film is above room temperature and its CO state is quite stable even at a high magnetic field of 12 T (Fig. 5.24**c**). On the other hand, for PCSMO film with T_{CO} around 205 K, the CO state collapses when the applied field exceeds 2.5 T, and a subsequent insulator-metal (I-M) transition occurs.

At the interface of the manganite/STON junctions, depletion layers are formed due to the diffusion of carries between manganites and STON [5, 13]. The removal of free charge carriers in a depletion layer with a thickness of Wd (Fig. 5.24**a**), leaving none to carry a current, results in an excellent insulator region working as a coupling capacitor to manganites (Fig. 5.24**b**). The carrier density of manganite films ($n_p \sim 10^{22}$ cm^{-3}) is much higher than that in STON ($n_n \sim 10^{17}$ cm^{-3}) and then $\varepsilon_p n_p \gg \varepsilon_n n_n$ in which $\varepsilon_p(\varepsilon_n)$ are the static dielectric constants in $p(n)$-layer. As a consequence, the depletion layers of these junctions are almost all located at the STON side according to the abrupt junction model [13]. In Fig. 5.24**d**, the current density-voltage ($J - V$) relation of junctions show good rectification, indicating the good insulator behavior of depletion layer in STON. Following this scenario, the built-in potentials (qV_{bi}) of these junctions were derived from $1/C^2 - V$ curves as shown in the Fig. 5.24**d**; 0.88, 0.83, and 0.38 eV for PCMO, PCSMO, and LSMO samples, respectively.

In the junction geometry, the manganite films are arranged in series with a capacitance of junction C_j. As schematically shown in Fig. 5.24**b**, the metallic LSMO film plays a role of a series resistor in the circuit. For PCMO and PC-SMO cases, however, the manganite films can be modeled as a capacitor (C_f) with a parallel resistor (R_f), both connected in series to larger C_j. With this

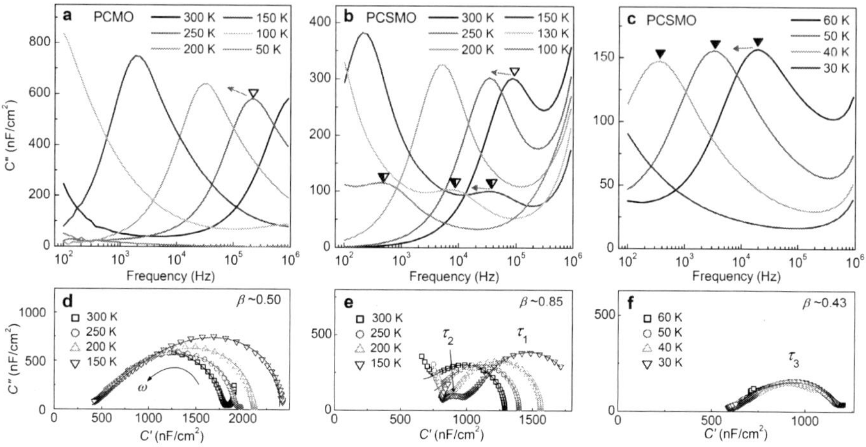

Figure 5.25. (**a-c**) Frequency dependence of imaginary (C'') part of complex capacitances at various temperatures for manganite films; (**a**) PCMO, (**b**) and (**c**) PCSMO. Blank, half-filled, and filled triangles indicate different loss peaks of C'' (relaxation τ_1, τ_2, and τ_3, respectively). Red arrows indicate the movement of loss peaks with decreasing temperature. (**d-f**) The corresponding Cole-Cole plots (C' vs. C''); (**d**) PCMO, (**e**) and (**f**) PCSMO. The solid lines are fitting results according to the Davidson-Cole model. The Davidson-Cole parameter β obtained from fitting results is $\sim$0.5 for PCMO. For PCSMO, it is about 0.85 for relaxation τ_1, about 1 for relaxation τ_2, and as small as 0.43 for relaxation τ_3. The labels τ_1, τ_2, and τ_3 in e and f indicate the corresponding arcs in Cole-Cole plots for different relaxation τ_1, τ_2, and τ_3, respectively (Reprinted by permission from Macmillan Publishers Ltd: Nature Communications (Ref. [28]), Copyright (2012)).

configuration, the dielectric properties of manganite films can be elucidated by analyzing the dielectric properties of the whole junction setup.

The frequency-dependent complex capacitance $C^*(\omega)$ has been measured for three samples with a capacitive technique at temperatures between 8 and 300 K. The complex dielectric spectrum is expressed as $C^*(\omega) = C'(\omega) - iC''(\omega)$, where $C'(\omega)$ and $C''(\omega)$ are the real and imaginary parts of the capacitance, respectively. Typical results for PCMO and PCSMO samples are shown in Fig. 5.25. For the metallic LSMO sample, dielectric response is dominated by C_j of the depletion layer in STON and has nearly no frequency dependence in a whole frequency window ($10^2 \sim 10^6$ Hz). In contrast to the LSMO case, dielectric relaxation behaviors are clearly observed in both PCMO and PCSMO samples. The $C''(\omega)$ has a loss peak at a specific frequency, which shifts exponentially to higher frequencies with increasing temperature; this is the characteristic of Debye-like relaxation with a thermally activated relaxation rate. The frequency of the loss peak (f_p) is related to a relaxation time as $\tau_p = 1/2\pi f_p \equiv 1/\omega_p$.

In PCMO sample with the CO state, only one kind of loss peak can be discerned in our frequency window as marked by blank triangles in Fig. 5.25**a**. In PCSMO, however, the situation is quite different. As shown in Fig. 5.25**b**, there is a discerned loss peak (defined as τ_1) at high temperatures and the peak position moves to the low f region when the temperature decreases. Below 170 K, another weaker loss peak (defined as τ_2) emerges at higher f region as marked by half-filled triangles in Fig. 5.25**b**. With further decreasing temperature, the third relaxation τ_3 (marked by filled triangles) comes to the present measurement window as shown in Fig. 5.25**c**. Similarly to τ_1, the peak positions of τ_2 and τ_3 shift to lower f region with decreasing temperature. It is interesting to find that the magnitude of loss peaks τ_3 is higher than that of τ_2 and smaller than that of τ_1. The main loss peak τ_1 corresponds to the relaxation process of the main phase in PCSMO sample and its magnitude is similar to that of PCMO sample. On the contrary, the additional relaxations, τ_2 and τ_3, originate from the phase-separation of the manganite film, will be discussed in detail later.

Figure 5.25**d-f** present the corresponding Cole-Cole plots ($C' vs. C''$) of PCMO and PCSMO samples. These experimental data provide a clear evidence that the complex capacitance spectra $C^*(\omega)$ of these two samples largely deviate from the simple Debye model (half circle). The physical processes of these deviations, manifested by broader and asymmetric shapes of the absorption term C'' of the spectra, are often described by Davidson-Cole model [54],

$$C^*(\omega) = C_{HF} + (C_{LF} - C_{HF})/[1 + i\omega\tau]^\beta \tag{5}$$

Here, $C_{HF}(C_{HF})$ is the high (low) frequency limit of the capacitance, $\omega = 2\pi f$ the angular frequency, τ the relaxation time, and β the Davidson-Cole parameter. All the $C' vs. C''$ curves can be well fitted with the Davidson-Cole model and the fitting results are shown by solid lines in Fig. 5.25. Interestingly, we found that the parameter β for a specific relaxation process is almost constant against temperature variation, whereas its values are different for different relaxations. It is around 0.5 for PCMO at all temperatures. For PCSMO, it is about 0.85 for relaxation τ_1, about 1 for relaxation τ_2, and as small as 0.43 for relaxation τ_3. The deviation of β from 1 indicates the distribution of multiple relaxation times in these strong-correlated oxides.

In the present junction configuration, the total capacitance C can be given approximately by $1/C$=$1/C^j$+$1/C^f$. Because of large dielectric constant in STON crystal especially at low temperatures ($\sim 10^3$ at 100 K) [55], C_j is much larger than C_f, therefore the contribution from C_f is dominant in the measured C value. As shown by Cole-Cole plots in Fig. 5.25**d-f**, the variation of C_{HF} and C_{LF} can reflect the change of dielectric properties of samples. In PCMO sample, as shown in Fig. 5.25**d**, the C_{LF} increases with decreasing of temperature while the C_{HF} is almost temperature independent. We simulated $C^*(\omega)$ results by using an equivalent circuit and Barrett's law for dielectric properties of STON depletion layer C_j [55]. The simulation results are consistent with this premise and the dielectric properties of manganite films could be deduced from the measured $C^*(\omega)$ spectroscopy. It was found that the static dielectric constants ε_r of PCMO and PCSMO films are 39 and 22 at room temperature, respectively. The ε_r of PCMO film is almost temperature independent, while the ε_r of main phase in PCSMO film increases with decreasing of temperature and it reaches around 27 at T=150 K, as derived from variation of both C_{HF} and C_{LF} (Fig. 5.25**e**). In Fig. 5.25**f**, on the other hand, both C_{HF} and C_{LF} are temperature-independent for the relaxation τ_3. It indicates the temperature insensitivity of dielectric strength of τ_3 which may be correlated to the ferromagnetic metal (FMM) phase involved in PCSMO film ($videinfra$).

Figure 5.26**a** and **b** show the external magnetic field effect on the dielectric responses of PCMO and PCSMO samples measured at 150 K. It is found that PCSMO shows a huge magneto-capacitive effect, while PCMO with the CO state is resistant to magnetic field. The relative magnetocapacitance $\Delta C/C$ of PCSMO can reach 46% with a magnetic field of 3 T. Moreover, the magnetic field can shift the main loss peak of $C''(\omega)$ for PCSMO film to higher frequency region which corresponds to a tendency of a faster relaxation, e.g., the peak f_p (τ_p) from 2.7×10^2 Hz (6.7×10^{-4} s) to 4.8×10^5 Hz (3.3×10^{-7} s) upon application of 9 T field. The field sensitive/insensitive dielectric relaxations in PCSMO and PCMO are similar to the respective magnetoresistance features as shown in Fig. 5.26**c** and **d**. Recently, it was reported that the dc magnetoresistance with help from Maxwell-Wagner (MW) effect could contribute to the magnetocapacitance especially in the materials with heterogeneous components. In this junction configuration, the resistance of a manganite film with 40 nm ($\sim 10^{-7}$ Ω) is quite smaller than the value which can be resonant with capacitance of the depletion layer (C_j $\sim 10^{-10}$F) in measurement window. Moreover, the

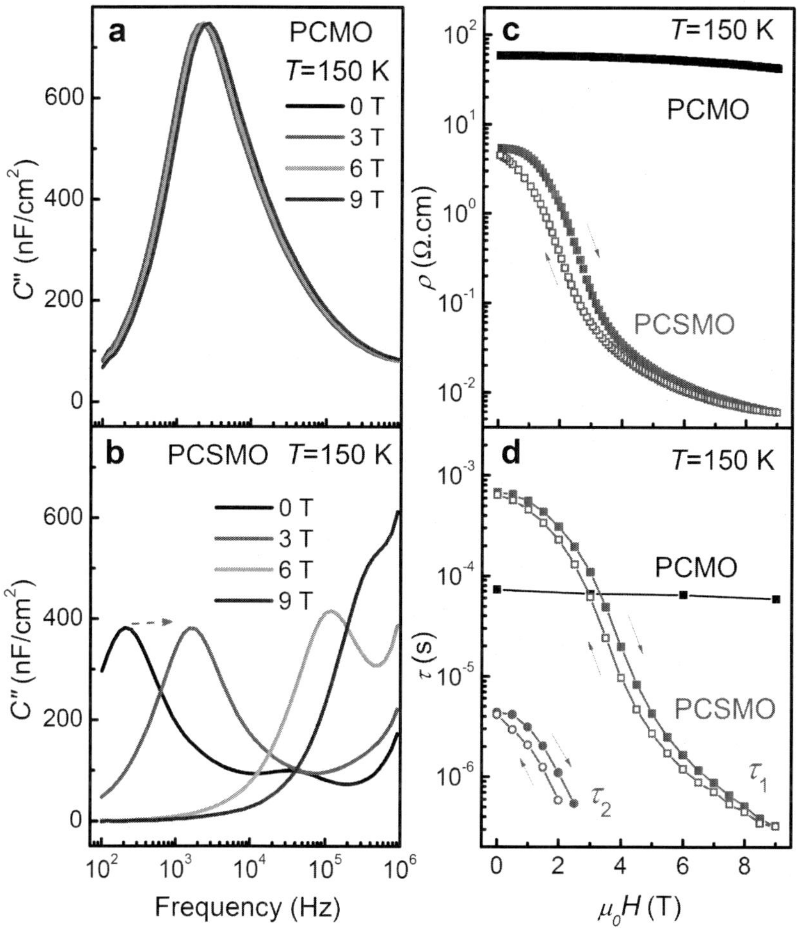

Figure 5.26. Frequency dependence of C'' under various magnetic fields for (**a**) PCMO and (**b**) PCSMO samples measured at T=150 K. Red dashed arrows in b indicate the movement of loss peaks with increasing magnetic field. (**c**) Magnetic field dependence of resistivity for PCMO (black) and PCSMO (red) films measured at T=150 K. (**d**) Magnetic field dependence of relaxation time τ for PCMO (black) and PCSMO (red) junctions measured at T=150 K. The red curves in d with squares and circles present results for relaxation τ_1 and τ_2 of PCSMO junction, respectively. In both **c** and **d**, the red arrows indicate the scan direction of magnetic field (Reprinted by permission from Macmillan Publishers Ltd: Nature Communications (Ref. [28]), Copyright (2012)).

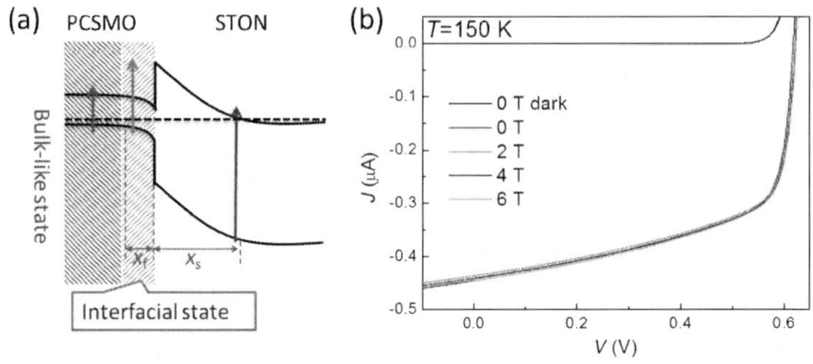

Figure 5.27. (a)Schematic image of band profiel for $Pr_{0.55}(Ca_{1-y}Sr_y)_{0.45}MnO_3$ (PCSMO)/STON heterojunction. (b)The current density versus voltage ($J - V$) relations for PCSMO/STON junction measured at 150 K. The magnetic fields were applied parallel to the film plane during the $J - V$ measurements.(unpublished data from the author)

asymmetric relaxation behavior (β<1)is not favorable for a simple MW model. Therefore, the suitable equivalent circuit for our junctions should be much more complicated if the dielectric relaxations observed here originate from the MW effect. Another possible origin for these dielectric relaxations is the contribution from small polarons in manganite films. From this viewpoint, the similar temperature/magnetic field dependences for relaxation time τ and dc resistance observed here indicate the dominant role of polaronic charge carriers in both ac dielectric and dc electrical transport properties.

After several rounds of testing, however, we found that the realization of a manganite-based heterojucntion with a correlated interface is challenging. That is because the physical properties of correlated materials at the interface, in general, deviate from that of bulk due to the breaking of symmetry or epitaxial strain effect [56, 57]. The magnetic field effect on the PVE of PCSMO/STON junctions have been studied and the typical results were shown in Fig.5.27. It was found that the magnetic field effect on the PVE is negligible even up to 6 Tesla. This indicates the interfacial state of PCSMO film is magnetic field insensitive and it deviates away from the bulk property. As mentioned above however, a interface with a clear first-order phase transition (IMT) is necessary to realize the multiple carrier generation and other controllable properties. Therefore, to control the physical property of interfacial layer plays a key role in this project.

From the viewpoint of interfacial engineering, the controlling of lattice distortion is very important in manganites films and heterojunctions. For PCSMO films (a=3.829 Åfor pseudo-cubic coordination) on STON substrate (a=3.905 Å), the lattice mismatch [defined as $(a_s - a_f)/a_s$] is as large as 1.9%. To reduce the lattice mismatch, we can change the compounds either PCSMO or STON. The former one is convenient and operable. From tens of candidates, lanthanum strontium manganese oxide ($La_{0.7}Sr_{0.3}MnO_3$, LSMO) with a=3.876 Åwas selected carefully. For LSMO case, the lattice mismatch between film and STON substrate has been reduced down to ~0.74 %, which is less than half of that for PCSMO case.

Both (001) and (110) oriented substrates were chosen for the preparation of LSMO/STON junctions. X-ray reciprocal space mapping (RSM) was used to study the lattice structure of the LSMO films as shown in Fig.5.28. The LSMO (a_{bulk} ~3.876 Å) films were subject to an in-plane tensile strain from STO (a ~3.905 Å) due to the original lattice mismatch. In all RSMs, the Q_x values of the peaks coincide for film and substrate, which indicates that both (001) and (110) films are epitaxially grown and the in-plane lattices are coherently connected to those of the STO substrate. In Fig. 5.28**a**, the Q_z values for the (-103) and (103) reflection peaks are the same, which indicates a tetragonal distortion for the (001) film. As for the (110) film, the Q_z values for the (310) and (130) reflection peaks are the same, while those for the (22-2) and (222) reflection peaks are deviated from each other. These features indicate the tilting of the LSMO (110) unit cell as shown in the schematics to the right. The difference between off-axis lattices [222] and [22-2] is around 0.83% and the tilting angle (90°-α) along the [001] direction is revealed as 0.5^o for this sample [56]. Examinations of RSM on various points of the film indicate that the tilting structure forms a single domain across the whole sample.

The tetragonal distortion of LSMO (001) film and monoclinic distortion of (110) film have been depicted in Fig. 5.29**a** and **b**, respectively. From the X-ray diffraction results, it was found that when the film thickness (t) is less than a critical value (approximately 40 nm), the unit-cell volume of the LSMO films expands beyond the bulk value owing to the biaxial in-plain strain from the substrate. The expansion for the (110) film (0.98% with t = 24 nm) is larger than that for the (001) film (0.89%) and this tendency becomes prominent near the interface [59]. Figure 5.29**d** shows the resistivity for both the (001) and the (110) films with various thicknesses as functions of temperature and magnetic field. In the comparision of the (110) and (001) films at the same thickness,

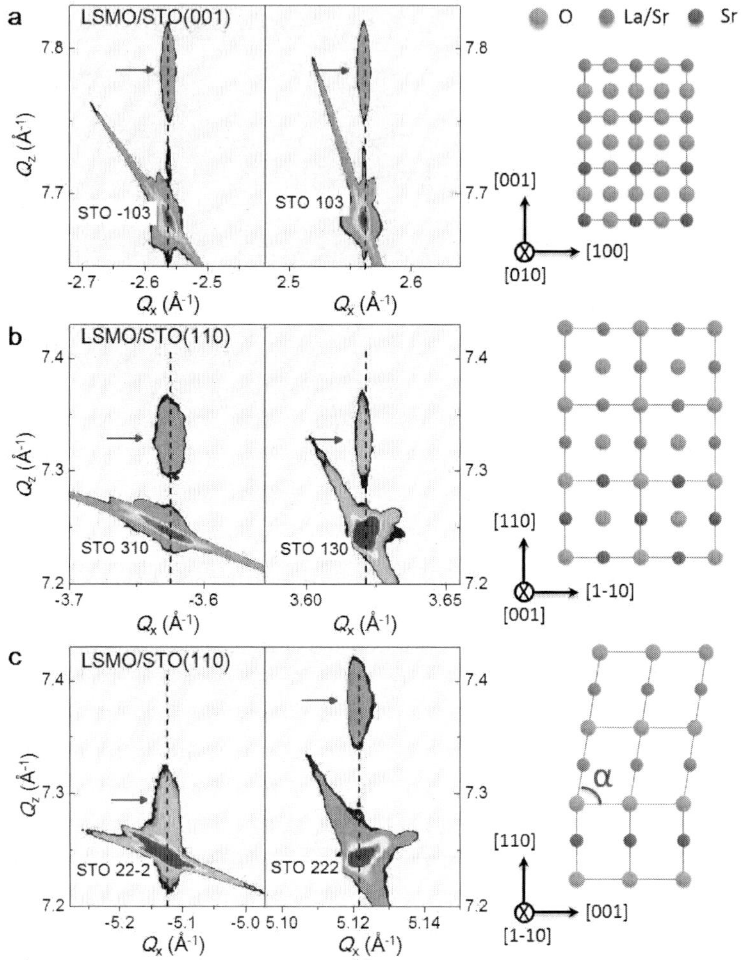

Figure 5.28. **a**, RSMs of a 24-nm-thick LSMO (001) film taken around the (-103) and (103) reflections. **b**, RSMs of a 24-nm-thick LSMO (110) film taken around the (310) and (130) reflections. **c**, RSMs of LSMO (110) film taken around the (22-2) and (222) reflections. The arrows indicate the positions of film peaks. The corresponding lattice structures for each film are schematically shown on the right (Reprinted by permission from Macmillan Publishers Ltd: Nature Communications (Ref. [47]), Copyright (2014)).

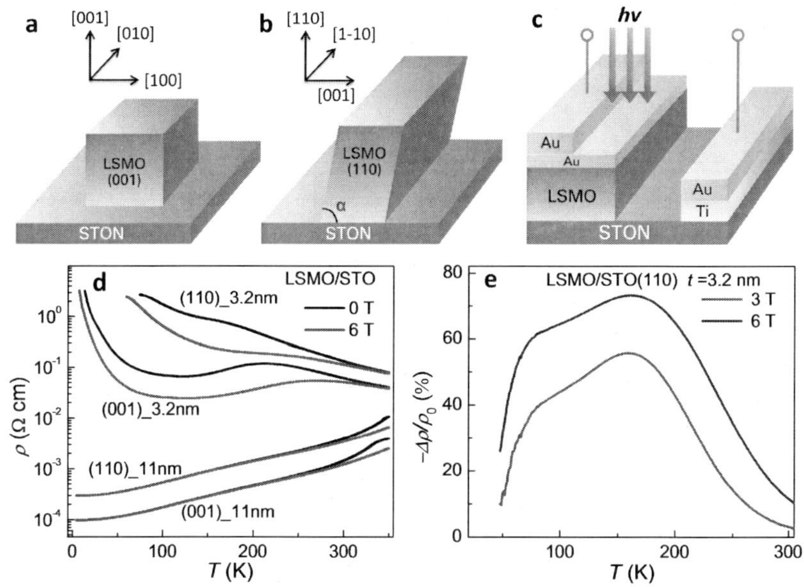

Figure 5.29. Crystal orientation and structure of **a** $La_{0.7}Sr_{0.3}MnO_3$ (LSMO) (001) and **b** LSMO (110) films on Nb-doped $SrTiO_3$ (STON) substrates. **c**, Device structure for the photovoltaic effect measurements. **d**, Temperature dependence of the resistivity for LSMO (001) and (110) films with different thicknesses deposited on Nb-free $SrTiO_3$. The negative magnetoresistance is more pronounced for thinner films. **e**, The relative change in resistivity ($-\Delta\rho/\rho_0$) as a function of temperature for the 3.2-nm-thick LSMO (110) film under 3- and 6-T fields (Reprinted by permission from Macmillan Publishers Ltd: Nature Communications (Ref. [47]), Copyright (2014)).

the tilt-distorted (110) films with larger cell-volume expansion exhibit higher resistivity than the (001) films, a trend more prominent in thinner films. For instance, the (110) film with t = 3.2 nm shows a semiconductor-like transport behavior even under a magnetic field of 6 T, while an insulator-metal transition is observed in the (001) film with the same thickness. The higher resistivity in (110) films may relate to the shear mode lattice strain or to the expansion of the cell volume [56, 59], but clarifying the detailed mechanism need further studying. On the whole, it is plausible to consider that the (001) interfaces are more conductive than those in the (110) case. The temperature dependence of magnetoresistance (-$\Delta\rho/\rho_0$) for the 3.2-nm-thick (110) film is shown in Fig. 5.29**e**, indicating a peak around 175 K for both 3- and 6-T fields.

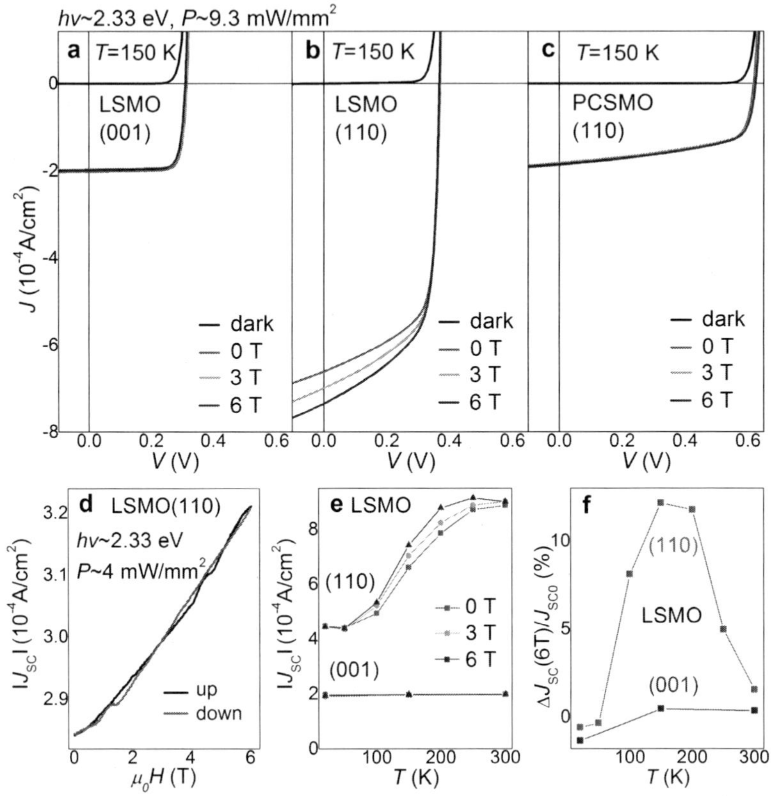

Figure 5.30. **a**, **b**, and **c** show current density versus voltage ($J-V$) relations for (001), (110), and PCSMO junctions, respectively. **d** shows the magnetic-field dependence of J_{SC} for (110) junction with an illumination intensity P = 4 mW/mm^2. **e**, Temperature dependence of J_{SC} for both (001) and (110) junctions. f, Temperature dependence of $\Delta J_{SC}(\mu_0 H$=6 T)/J_{SC}(0T) for these junctions (Reprinted by permission from Macmillan Publishers Ltd: Nature Communications (Ref. [47]), Copyright (2014)).

By combining an LSMO film (t = 24 nm) and an n-type Nb-doped (0.01 wt%) $SrTiO_3$ (STON) substrate, rectifying heterojunctions were formed. The current density versus voltage ($J-V$) characteristics of the (001) and (110) junctions were measured with an illumination from a laser diode ($h\nu \sim 2.33$

eV) under a magnetic field parallel to the film plane. The schematics of the junction device and experimental setup are shown in Fig. 5.29**c**. In dark conditions, both junctions have excellent rectification with $J-V$ curves crossing the origin as shown in Fig. 5.30**a** and 5.30**b**. When the junctions are illuminated, the $J-V$ curves are shifted downward by the photocurrent. The illumination intensity is set at $P = 9.3$ mW/mm^2 which is a decade higher than a typical solar intensity (1mW/mm^2). Note the following two points from the comparison of Figs. 5.30**a** and 5.30**b**. First, the J_{SC} for the (110) junction is larger than that of the (001) junction by a factor of three. Since the incident photon energy used here is lower than the band gap energy of STON ($E_{gs} \sim 3.2$ eV), only the excited photocarriers on the LSMO side contribute to the photocurrent. Therefore, the greater photocurrent in the (110) junction indicates a wider active interfacial layer than in the (001) junction. Furthermore, the magnetic-field effect on the photocurrent is negligibly small in the (001) junction, while it is of significant magnitude in the (110) case. The J_{SC} is enhanced by 12% through the application of 6 T at 150 K; and this field-induced enhancement is almost linear with applied magnetic field, as shown in Fig. 5.30**d**. Moreover, the magnetic-field effect in the (110) junction is most prominent in the middle temperature range (Fig. 5.30**e**), and the relative change under 6 T, $\Delta J_{SC}/J_{SC0}$, has a peak around 150-200 K (Fig. 5.30**f**). This temperature dependence is similar to that observed in magnetoresistance (Fig. 5.29**e**). For the (001) junction, on the other hand, the magnetic-field effect is hardly observed across the entire temperature measurement range. As another reference, the results of PCSMO/STON junction are shown in Fig. 5.30**c**, which is similar to the behavior of (001) case.

To gain a deeper insight into the magneto-tunability in LSMO/STON junctions, the magnetic-field effect on the $J-V$ characteristics, and the junction capacitance (C) of LSMO(001), LSMO(110), and PCSMO junctions were investigated in dark conditions as shown in Fig. 5.31. While all junctions show a highly rectifying diode property, the (110) junction has a relatively greater reverse saturation current density (J_S) than does the (001) junction, which is consistent with previous results in $La_{0.6}Sr_{0.4}MnO_3$/STON junctions [58]. In addition, J_S in the (110) junction can be modulated by the magnetic field, whereas there is no magnetic-field effect within the $J-V$ range of the (001) junction. Figures 5.31**d** and 5.31**e** display the $1/C^2-V$ relationships of the (001) and (110) junctions, respectively. Fitting of $1/C^2-V$ curves with linear lines and their extrapolations onto the voltage axis (dashed lines) give the built-in potential (V_{bi}) of the junctions, which are 0.50 V, 0.58 V, and 0.98 for the (001),

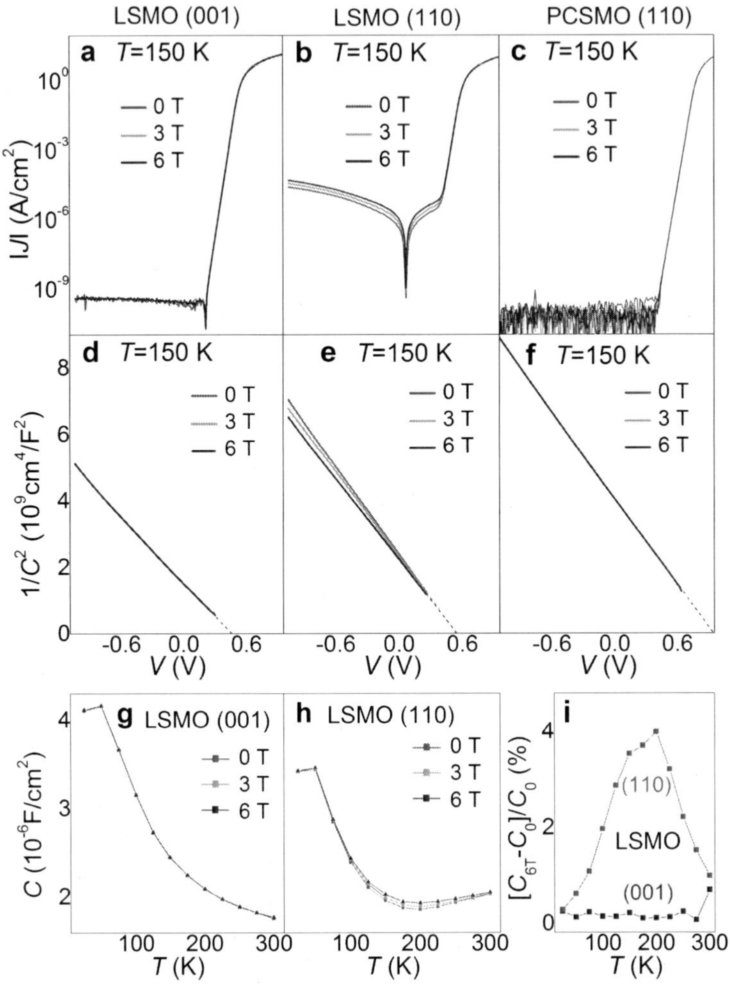

Figure 5.31. **a**, **b**, and **c** are the $J - V$ curves for (001), (110), and PCSMO junctions, respectively. **d**, **e**, and **f** show the $1/C^2 - V$ relationships for (001), (110), and PCSMO junctions, respectively. The dashed lines are the results of linear fitting. **g**, and **h** are the temperature dependence of C at zero bias for LSMO(001) and (110) junctions, respectively. **i**, The temperature dependence of relative capacitance change $[C_{6T} - C_0]/C_0$ for LSMO(001) and (110) junctions (Reprinted by permission from Macmillan Publishers Ltd: Nature Communications (Ref. [47]), Copyright (2014)).

(110), and PCSMO junctions, respectively. Similar to the $J-V$ relationships, the magnetic-field effect on C is negligibly small in the (001) junction across the bias voltage (Fig. 3d) and temperature (Fig. 3g) ranges, which is consistent with previous reports [58]. For the (110) junction, on the other hand, C is clearly enhanced by the magnetic field (Fig. 5.31**e** and 5.31**h**). We note that while the V_{bi} of the (110) junction is not changed by the application of a magnetic field, the gradient of the $1/C^2 - V$ curve decreases with increasing magnetic field. Again, this field induced enhancement is temperature dependent, and the relative change $[C_{6T} - C_0]/C_0$ has a peak around 150-200 K as shown in Fig. 5.31**i**.

Although magnetic-field effects were observed in both $J-V$ and $C-V$ curves of the (110) junction, we consider their origins to be different. Numerical simulations have been performed for a semiconductor-based junction model taking shunt resistance (R_{SH}) into account [13]. The magnetic-field dependence of the $J-V$ curve for the (110) junction in dark conditions can be well reproduced by changing R_{SH} in simulation results. On the other hand, the same amount of change in R_{SH} has negligible effect on C and photocurrent. These results indicate that the origin of greater J_S in the (110) junction, and its change under magnetic fields, can be assigned to the shunt resistance, but the magnetic-field effects on J_{SC} and C must originate from other factors.

The photocurrent density (J_{ph}) is known to have a close relationship with the width of the depletion layers since $J_{ph} \sim qG(W_f + W_s)$, where G is the generation rate of photoexcited electron-hole pairs, W_f and W_s are the depletion-layer widths in LSMO and STON, respectively. In the constituent materials, the STON is magnetically inactive, while the magnetic field can modulate the electronic structure of LSMO, G, and W_f due to the strong correlation. The absence of the magnetic-field effect on J_{ph} in the LSMO (001) junction indicates that W_f is negligibly small or constant under magnetic fields. Considering that the interfacial layer in the LSMO (001) junction is rather metallic (Fig. 5.29**d**), W_f will be negligible and thus we can regard it as a Schottky junction, shown schematically in the left inset of Fig. 5.32**e**, which is consistent with previous reports [60, 61]. For the (110) junction with tilt-distorted LSMO film, in contrast, a more insulating interfacial layer (5.29**d**) and magneto-tunable J_{ph} and C signify the existence of a finite W_f in the LSMO (110) side. Hence, the band profile of the (110) junction is rather close to the pn junction as schematically shown in the two right insets of Fig. 5.32**e**.

The above scenarios on the interfacial band profiles and their magnetic-field

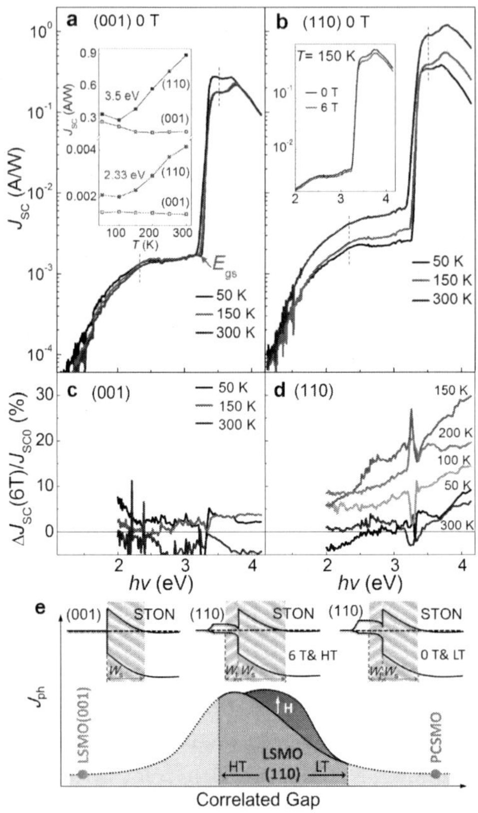

Figure 5.32. **a** and **b** show the photocurrent spectra measured at various temperatures without magnetic field for (001) and (110) junctions, respectively. The inset of **a** shows the temperature dependence of J_{SC} at selected photon energies of 2.33 and 3.5 eV (marked by dashed vertical lines in a and b). The upper inset of **b** shows photocurrent spectra under 0 T and 6-T magnetic fields taken at 150 K. **c** and **d** display the relative photocurrent change ($\Delta J_{SC}/J_{SC0}$ with the application of a 6-T field as functions of photon energy and temperature for LSMO(001) and (110) junctions, respectively (Due to poor signal-to-noise ratio, data obtained below $h\nu$ = 2 eV are omitted here). **e**, A schematic image of the relationship between J_{SC} and correlated gap in correlated junctions. The insets show the schematic band diagrams for (left) LSMO (001) junctions, (middle) LSMO (110) junction under high magnetic field (6 T) or high temperature (HT), and (right) LSMO (110) junction without magnetic field (0 T) or low temperature (LT), respectively (Reprinted by permission from Macmillan Publishers Ltd: Nature Communications (Ref. [47]), Copyright (2014)).

effects are also supported by the photocurrent action spectra (J_{SC} versus $h\nu$) as shown in Fig. 5.32. For both (001) and (110) junctions, the J_{SC} has a jump at around 3.2 eV (marked by an arrow in Fig. 5.32**a**), which corresponds to the E_{gs} of STON. The J_{SC} above (below) E_{gs} originates mainly from the photocarriers generated in the depletion layer of STON (LSMO). To draw further information on the depletion layer in each region, we plot the temperature dependence of J_{SC} for photon energies below (hv = 2.33 eV) and above (3.5 eV) E_{gs} in the inset of Fig. 5.32**a**. Below E_{gs}, J_{SC} is almost temperature independent in the (001) junction. Above E_{gs}, J_{SC} increases slowly at first with cooling and then increases rapidly below the temperatures of 150 K. This feature reflects the variation of Ws in STON ($W_s \sim \varepsilon_s^{1/2}$), as is evident in the capacitance ($C_j \sim \varepsilon_s/W_s \sim \varepsilon_s^{1/2}$) results shown in Fig. 5.31**g**. Moreover, it is found that the magnetic-field effect on J_{ph} of the (001) junction is negligibly small across the whole spectrum (Fig. 5.32**c**) and is consistent with the J_{ph}-V curves (Fig. 5.30**a**).

As for the (110) $p-n$ junction with finite W_f, the situation is quite different, where the temperature/magnetic-field dependent correlated gap (E_{gf}) plays an important role. In the correlated electron manganite, "Mottness" is controlled by external fields [51, 53], i.e., the temperature/magnetic-field dependence of the insulating gap in the LSMO interfacial layer is expected (the right two insets of Fig. 5.32**e**). An increase in temperature or magnetic field brings out shrinkage of E_{gf} and an enhancement of the effective carrier density (n_f). As a result, on the one hand, the increase of n_f can enhance W_s ($\sim [(\varepsilon_s\varepsilon_f n_f)/n_s/(\varepsilon_s n_s + \varepsilon_f n_f)]^{1/2}$) on the STON side proportionally, while decreases W_f ($\sim [(\varepsilon_s\varepsilon_f n_s)/n_f/(\varepsilon_s n_s + \varepsilon_f n_f)]^{1/2}$) on the LSMO side [13], n_s being the carrier density of substrate, and ε_f the dielectric constant of the LSMO film. On the other hand, in accord with the shrinkage of E_{gf}, the increase in temperature or magnetic field drives this correlated system closer to the phase boundary with metal. Then, the photocurrent, which is related to the phase stability, may be enhanced [52]. For the $h\nu < E_{gs}$ region, obviously, the latter factor is dominant, and higher temperature or higher magnetic field produces the larger J_{SC}, as shown in Fig. 5.30**e** and the inset of Fig. 5.32**a**. For the $h\nu > E_{gs}$ region, the former factor, i.e., W_s on the STON side, is dominant for J_{SC}. Then the enhancement of J_{SC} can also be produced with an increase of temperature and magnetic field, as shown in Fig. 5.32**b** and Fig. 5.32**d**. The $\Delta J_{SC}/J_{SC0}$ under 6 T (Fig. 5.32**d**) can reach up to approximately 30% with $h\nu = 4.1$ eV at 150 K.

These results indicate that the correlated gap E_{gf} plays a decisive role in the magnitude of J_{ph} and its temperature- and magneto-tunability in the correlated junction. This remarkable feature is summarized schematically in Fig. 5.32**e**. Zero E_{gf} (e.g., the (001) junction) with absence of W_f results in a small J_{ph} and negligible temperature/magnetic-field effects. The (110) junction with medium E_{gf} can produce a greater J_{ph} at high temperature or high magnetic field (left and up arrows in Fig. 5.32**e**) and its value will be reduced again when the E_{gf} increases further by lowering the temperature (right arrow) or removing the magnetic field. From this tendency, one can expect that a further increase of E_{gf}, accompanying a robust insulating state against external fields, will result in smaller J_{ph} as well as smaller temperature and magnetic-field effects. The results on the PCSMO junction with larger E_{gf} verify this expectation (Fig. 5.30**c**). The PCSMO is in highly robust insulating state at the interface and insensitive to small external magnetic fields used here. The most surprising and important finding in this study is that when the interface is in phase competing state (middle correlated gap) and the magneto-photocurrent effect is remarkable as in the LSMO (110) junction, the absolute magnitude of the photocurrent is simultaneously enhanced. This means that there is an optimum window of electron correlation for the effective harvest of photocurrent (Fig. 5.32**e**). One possible origin for the photocurrent enhancement we have in mind is the local phase transition from the gapped to metallic state near the interface. When the interface is in slightly gap-opened state and keen phase competition, it becomes most amenable to the photoexcitation.

The results shown above evidence that small structural and chemical changes to perovskite oxides can induce critical interfacial phase competition in these materials. It reveals that the choices of the appropriate substrate orientation as well as the chemical composition are critical factors to tune the correlated gap and approach the optimum window. while this technology is at an early stage, it can guide development of next-generation of photovoltaic devices.

References

[1] Janice Nickel, "Magnetoresistance Overview", *HP HEWLETT PACKARD* 1995.

[2] C. Zener, *Phys. Rev.* 81, 440 (1951); *ibid* 82, 403 (1951).

[3] Zhigao Sheng, Yuping Sun, Xuebin Zhu, Wenhai Song, and Peng Yan, *J. Phys. D: Appl. Phys.* 40, 3300-3305(2007).

[4] C. Mitra, P. Raychaudhuri, K. Dörr, K. -H. Müller, L. Schultz, P. M. Oppencer, and S. Wirth, *Phys. Rev. Lett.* 90. 017202 (2003).

[5] H. Tanaka, J. Zhang, and T. Kawai, *Phys. Rev. Lett.* 88, 027204 (2002).

[6] H. Li, J. R. Sun, and H. K. Wong, *Appl. Phys. Lett.* 80, 628 (2002).

[7] M. Bowen, M. Bibes, A. Barthelemy, J, P. Contour, A. Anane, Y. Lemaire, and A. Fert, *Appl. Phys. Lett.* 82, 233 (2003).

[8] K. Ghosh, S. B. Ogale, S. P. Pai, M. Robson, Eric Li, I. Jin, Zi-wen Dong, R. L. Greene, and M. Johnson, *Appl. Phys. Lett.* 74, 689 (1998).

[9] Z. G. Sheng, Y. P. Sun, J. M. Dai, X. B. Zhu, W. H. Song, and Z. R. Yang, *J. Phys. D: Appl. Phys.* 42, 085002(2009).

[10] J. R. Sun, C. M. Xiong, T. Y. Zhao, S. Y. Zhang, Y. F. Chen, and B. G. Shen, *Appl. Phys. Lett.* 84, 1528 (2004).

[11] J. R. Sun, C. M. Xiong, and B. G. Shen, *Appl. Phys. Lett.* 85, 4977 (2004).

[12] Z. G. Sheng, W. H. Song, Y. P. Sun, J. R. Sun, and B. G. Shen, *Appl. Phys. Lett.* 87, 032501 (2005).

[13] S. M. Sze, Physics of Semiconductor Devices, 2nd ed. (Wiley, New York, 1981).

[14] W. Eerenstein, M. Wiora, J. L. Prieto, J. F. Scott, N. D. Mathur, *Nature Materials* 6, 348 (2007).

[15] J. M. D. Coey, M. Viret, S. von Molnár, *Advances in Physics* 48, 167 (1999) and references therein.

[16] I. Pallecchi, L. Pellegrino, E. Bellingeri, A. S. Siri, D. Marré, A. Tebano and G. Balestrino, *Phys. Rev. B* 78, 024411(2008).

[17] X. Hong, A. Posadas, and C. H. Ahn, *Appl. Phys. Lett.* 86, 142501 (2005).

[18] T. Wu, S. B. Ogale, J. E. Garrison, B. Nagaraj, Amlan Biswas, Z. Chen, R. L. Greene, R. Ramesh, T. Venkatesan, and A. J. Millis, *Phys. Rev. Lett.* 86, 5998 (2001).

[19] B. Wiedenhorst, C. Hofener, Y. F. Lu, J. Klein, L. Alff, R. Gross, B. H. Freitag, W. Mader, *Appl. Phys. Lett.* 74, 3636(1999).

[20] M. Angeloni, G. Balestrino, N. G. Boggio, P. G. Medaglia, P. Orgiani, and A. Tebano, *J. Appl. Phys.* 96, 6387 (2004).

[21] A. J. Millis, T. Darling, and A. Migliori, *J. Appl. Phys.* 83, 1588 (1998).

[22] Guanyin Gao, Shaowei Jin, and Wenbin Wu, *Appl. Phys. Lett.* 90, 012509 (2007).

[23] C. Thiele, K. Dörr, S. Fähler, and L. Schultz, D. C. Meyer, A. A. Levin, and P. Paufler, *Appl. Phys. Lett.* 87, 262502 (2005).

[24] C. Thiele, K. Dörr, O. Bilani, J. Rodel, and L. Schultz, *Phys. Rev. B* 75, 054408 (2007).

[25] R. K. Zheng, Y. Yang, H. L. W. Chan, C. L. Choy, and H. S. Luo, *Phys. Rev. B* 75, 212102 (2007).

[26] Z. G. Sheng, J. Gao, and Y. P. Sun, *Phys. Rev. B* 79, 174437(2009).

[27] T. W. Darling, A. Migliori, E. G. Moshopoulou, Stuart A. Trugman, J. J. Neumeier, J. L. Sarrao, A. R. Bishop, and J. D. Thompson, *Phys. Rev. B* 57,5093 (1998).

[28] Z. G. Sheng, M. Nakamura, F. Kagawa, M. Kawasaki, and Y. Tokura, *Nature Communications* 3, 944 (2012).

[29] Liuwan Zhang, Casey Israel. Amlan Biswas, R. L.Greene, and Alex de Lozanne, *Science* 298, 805(2002).

[30] Y. Tokura and N. Nagaosa, *Science* 288, 462(2000).

[31] E. Dagotto, Takashi Hotta, and Adriana Moreo, *Phys. Rep.* 344, 1 (2001).

[32] Y. Okimoto, and Y. Tokura, *J. Supercond. Nov. Magn.* 13, 27(2000).

[33] Y. Tokura, *Physics Today* 56, 50(2003).

[34] A. Asamitsu, Y. Tomioka, H. Kuwahara, Y. Tokura, *Nature* 388, 50 (1997).

[35] M. Fiebig. K. Miyano. Y. Tomioka. Y. Tokura, *Science* 286, 1925 (1998).

[36] R. Cauro, A. Gilabert, J. P. Contour, R. Lyonnet, M. -G. Medici, J. -C. Grenet, C. Leighton, and I. K. Shuller, *Phys. Rev. B* 63, 174423 (2001).

[37] R. Cauro, J. -C. Grenet, A. Gilabert, M. -G. Medici, *Int. J. Mod. Phys. B* 13, 3786 (1999).

[38] A. Gilabert, R. Cauro, M. G. Medici, J. C. Grenet, H. S. Wang, Y. F. Hu, and Qi Li, *J. Supercond. Nov. Magn.* 13, 285 (2000).

[39] Z. G. Sheng, J. Gao, and Y. P. Sun, *Phys. Rev. B* 79, 014433 (2009).

[40] J. Gao, S. Q. Shen, T. K. Li, and J. R. Sun, *Appl. Phys. Lett.* 82, 4732(2003).

[41] Y. G. Zhao, Y. H. Wang, G. M. Zhang, B. Zhang, X. P. Zhang, C. X. Yang, P. L. Lang, M. H. Zhu, and P. C. Guan, *Appl. Phys. Lett.* 86, 122502(2005).

[42] J. M. Dai, W, H, Song, S. G. Wang, S. L. Ye, K. Y. Wang, J. J. Du, and Y. P. Sun, *J. Appl. Phys.* 90, 3118 (2001).

[43] M. Fiebig, K. Miyano, Y. Tomioka, and Y. Tokura, *Science* 280, 1925 (1998).

[44] Z. G. Sheng, Y. P. Sun, X. B. Zhu, J. M. Dai, W. H. Song, and Z. R. Yang, *J. Appl. Phys.* 102, 093908 (2007).

[45] Z. G. Sheng, Y. P. Sun, J. M. Dai, X. B. Zhu, and W. H. Song, *Appl. Phys. Lett.* 89, 082503 (2006).

[46] Z. G. Sheng, B. C. Zhao, W. H. Song, Y. P. Sun, J. R. Sun, and B. G. Shen, *Appl. Phys. Lett.* 87, 242501 (2005).

[47] Z. G. Sheng, M. Nakamura, W. Koshibae, T. Makino, Y. Tokura and M. Kawasaki, *Nature Communications* 5, 4584 (2014).

[48] J. R. Sun, B. G. Shen, Z. G. Sheng, Y. P. Sun, *Appl. Phys. Lett.* 85, 3375(2004).

[49] Y. Tomioka, Y. Tokura, *Phys. Rev. B* 66, 104416 (2002).

[50] M. Matsubara, Y. Okimoto, T. Ogasawara, S. Iwai, Y. Tomioka, H. Okamoto, and Y. Tokura, *Phys. Rev. B* 77, 094410 (2008).

[51] W. Koshibae, N. Furukawa, N. Nagaosa, *Phys. Rev. Lett.* 103, 266402 (2009).

[52] W. Koshibae, N. Furukawa, N. Nagaosa, *Europhys. Lett.* 94, 27003 (2011).

[53] W. Koshibae, N. Furukawa, N. Nagaosa, *Phys. Rev. B* 87, 165126 (2013).

[54] D. W. Davidson, and R. H.Cole, *J. Chem. Phys.* 19, 1484-1490(1951).

[55] J. H. Barrett, *Phys. Rev.* 86, 118 (1952).

[56] H. Boschker, et al. *Phys. Rev. Lett.* 109, 157207 (2012).

[57] L. Fitting Kourkoutisa, J. H. Song, H. Y. Hwang, and D. A. Muller, *PNAS* 107 11682-11685 (2010).

[58] M. Minohara, Y. Furukawa, R. Yasuhara, H. Kumigashira, and M. Oshima, *Appl. Phys. Lett.* 94, 242106 (2009).

[59] R. V. Chopdekar, E. Arenholz, and Y. Suzuki, *Phys. Rev. B* 79, 104417 (2009).

[60] M. Nakamura, A. Sawa, H. Sato, H. Akoh, M. Kawasaki, Y. Tokura, *Phys. Rev. B* 75, 155103 (2007).

[61] M. Nakamura, A. Sawa, J. Fujioka, M. Kawasaki, Y. Tokura, *Phys. Rev. B* 82, 201101 (2010).

EDITORS' CONTACT INFORMATION

Dr. Zhigao Sheng
High Magnetic Field Laboratory,
Chinese Academy of Sciences
P.O. Box 1129, Hefei, Anhui 230031, P. R. China
Email: zhigaosheng@hmfl.com

Dr. Ling Hu
Key Laboratory of Materials Physics,
Institute of Solid State Physics,
Chinese Academy of Sciences,
P.O. Box 1129, Hefei, Anhui 230031, P. R. China
Email: huling@issp.ac.cn

Dr. Yuping Sun
Key Laboratory of Materials Physics,
Institute of Solid State Physics,
High Magnetic Field Laboratory,
Chinese Academy of Sciences
P.O. Box 1129, Hefei, Anhui 230031, P. R. China
Email: ypsun@issp.ac.cn

INDEX

A

B

C

D

E

F

G

H

I

K

L

M

S

T

U

V

W

X

Y